R Hodel

EQUIVALENTS OF THE AXIOM OF CHOICE

STUDIES IN LOGIC

AND

THE FOUNDATIONS OF MATHEMATICS

L. E. J. BROUWER
E. W. BETH
A. HEYTING

Editors

1970

NORTH-HOLLAND PUBLISHING COMPANY
AMSTERDAM – LONDON

EQUIVALENTS OF THE AXIOM OF CHOICE

HERMAN RUBIN
Professor of Statistics, Michigan State University

and

JEAN E. RUBIN
Lecturer in Mathematics and Philosophy
Michigan State University

1970

NORTH-HOLLAND PUBLISHING COMPANY
AMSTERDAM – LONDON

First Edition 1963
Second Printing 1970

PUBLISHERS:
NORTH-HOLLAND PUBLISHING COMPANY – AMSTERDAM
NORTH-HOLLAND PUBLISHING COMPANY LTD. – LONDON

ISBN 7204 2225 6

PRINTED IN THE NETHERLANDS

to
Arthur and Leonore

PREFACE

In 1955 at Stanford University Professor Patrick Suppes gave a course on axiomatic set theory, during which time Professor Herman Rubin was asked to give a series of lectures on the axiom of choice. During the course of these lectures it was noted that, while there was much in the literature on the axiom of choice, the material was available only in many diverse journals and books. It was suggested that we collect this material to make it moie readily available. It seemed like a simple enough project to begin with; then it grew and grew and grew some more and now has blossomed forth into a book.

The book consists of a selection of the forms of the axiom of choice which appeared in the literature together with additional forms which were obtained in the process of writing the book. It would have been a hopeless task to try to include all of the forms of the axiom of choice which appeared in the literature, so we chose the forms which in our opinion were either used often in practice, unusual, relatively unknown, or particularly weak or strong. We hope that we have included all of the interesting equivalents of the axiom of choice.

We assume a knowledge of logic and elementary set theory (von Neumann-Bernays-Gödel set theory), but we do include a list of definitions of set theoretical symbols and terms in the section entitled "Preliminary Definitions and Theorems".

In Part I we discuss propositions which are equivalent to the usual form of the axiom of choice. These equivalents will be referred to as *set forms*. In Part II we discuss stronger forms — essentially forms which are obtained from the set forms by changing the word "set" to "class". These latter forms are called *class forms*. The set forms of the axiom of choice are the forms which are most often used in practice.

In preparing this monograph for publication we first prepared a draft and sent it to several people for their comments and corrections. We are very grateful to the people who did reply. We believe that the quality and usefulness of the book was greatly improved by their comments. In particular we should like to thank Professor Alfred Tarski for his many useful comments and corrections. Others whom we should like to thank are Professors E. W. Beth, A. Levy, D. Scott, and R. Vaught. Our typists, Ann Breen and Barbara Johnson, also deserve credit for bearing with us under the strain, and we should like to thank the Mathematics and Statistics departments at Michigan State University for their cooperation.

HERMAN RUBIN
JEAN E. RUBIN

September 1961
East Lansing, Michigan

TABLE OF CONTENTS

INTRODUCTION

There has been quite a bit of philosophical discussion about whether or not the axiom of choice is meaningful. One form of the axiom of choice which was given by Russell [1] in 1906 and Zermelo [2] in 1908 is as follows:

If t is a disjoint collection of non-empty sets, there exists a set C which consists of one and only one element from each set in t.

The controversy is about the meaning of the word "exists". What does it mean for a set to exist? One group of mathematicians, loosely called the *intuitionists,* believes that a set exists only if each of its elements can be designated specifically or at least if there is a law by which each of its elements can be constructed. (The meaning of the word "constructible" varies greatly even among intuitionists. It should not be assumed that all intuitionists have the same criteria for the existence of a set.) So that for example if x is the set of even integers this set exists because there is a law to decide whether or not a given integer is in x. On the other hand, if y were defined to be the smallest natural number n which has the property that there is a sequence of 100-n digits all equal to each other in the decimal expansion of $(\pi + e)^{\pi - e}$, there is at present no rule to construct y, so that according to the intuitionists, the number y does not exist. (See Beth [2], pp. 418–19.) (There are some mathematicians who would accept the existence of the number y, but who do not accept the axiom of choice. Then there are others who not only do not accept the existence of the number y, but also do not accept the validity of an indirect proof.)

Another school of mathematicians takes a more liberal viewpoint about existence and they believe, essentially, that an axiom about existence of sets may be used if it does not lead to a contradiction.

It has been shown by Gödel [1] that the axiom of choice is consistent with the other axioms of set theory, (von Neumann-Bernays-Gödel set theory). Moreover, the axiom of choice has become almost indispensable in mathematics since a large number of important results have been obtained from it in almost all branches of mathematics without leading to a contradiction. So that the authors along with the other members of this school have no qualms about using the axiom of choice when necessary. However, there are some mathematicians who believe that the axiom of choice is obvious and they tend to use it in proofs whether it is necessary or not. Working on the principle that it is desirable to obtain the strongest possible results from the weakest possible hypotheses we recommend not using the axiom of choice unless it seems absolutely necessary. (Certain authors have replaced the axiom of choice with weaker propositions (like PR or PD below) in order to avoid using something as strong as the axiom of choice.)

Actually, however, whether the axiom of choice is realistic or not is immaterial for the purposes of this monograph. Our main purpose is to give many of the important propositions which are logically equivalent to the axiom of choice. That is, those propositions which can be proved equivalent to the axiom of choice using the axioms of set theory, without the axiom of choice, and the rules of logic. If the axiom of choice were not realistic then what we do here would still be logically correct, assuming our proofs are correct, even if it were not very useful.

Two important results of Fraenkel and Gödel have placed the question of the "truth" of the axiom of choice in almost the same position as that of the parallel postulate in geometry. In 1922 Fraenkel [1] proved that the axiom of choice is independent of the other axioms of set theory. That is, Fraenkel constructed a model in which the axioms of set theory excluding the axiom of choice are satisfied but this model contains a set which does not satisfy the axiom of choice. This result was later extended by Fraenkel [2], Lindenbaum and Mostowski [1], and Mostowski [1], [2], and [3].

However, these results are not completely satisfactory because the models used all contained individuals. In 1957, Specker [2] constructed a model in which there are no individuals but both the axiom of choice and the axiom of regularity are not satisfied. Mendelson [2] in 1956, obtained a little stronger result – a model with no individuals in which the axiom of regularity for finite sets is satisfied. The question of independence cannot be considered completely settled until a demonstration with no individuals and satisfying the axiom of regularity is given; in fact our forms P 17 and P 6S are satisfied in all the models with which we are acquainted.

As was mentioned above, Gödel [1] in 1940 proved that the axiom of choice is consistent with the other axioms of set theory (von Neumann-Bernays-Gödel set theory). He proved that given a model for set theory in which there are no individuals and the axiom of regularity is true, there exists a model in which, in addition, the axiom of choice is true. Moreover, if Gödel's model is modified so that either individuals exist or the axiom of regularity is false, the truth of the axiom of choice is not disturbed. Hence, the axiom of choice is consistent with the other axioms of set theory whether there exist individuals or not and whether the axiom of regularity is true or not. (These results do not apply to every proposed model of set theory. In fact, in Quine's model the axiom of choice is inconsistent. See Quine [1] and Specker [1].)

There are also some known results on the relative strength of the axiom of choice and some other well-known propositions. It was stated in 1926 in Lindenbaum and Tarski [1] that the generalized continuum hypothesis implies the axiom of choice (proven by Sierpenski [3] in 1947) and Tarski [4] has stated an axiom on inaccessible sets which implies the axiom of choice. (See also Bachman [1].) It is known that the generalized continuum hypothesis is consistent (Gödel [1]) but it is not known whether or not it is actually stronger than the axiom of choice. However, it is known that Tarski's axiom is stronger than the axiom of choice but it is not known if it is consistent. (Its negation is consistent with the axiom of choice and the other axioms of set theory.) The prime ideal theorem is known to follow from the axiom of choice but it wasn't until recently that Halpern [1] proved that the converse is not true. The axiom of choice is actually stronger than the prime ideal theorem. Also, the Birkhoff representation theorem for abstract algebras is known to follow from the axiom of choice, but it is not known if the converse is provable.

PRELIMINARY DEFINITIONS AND THEOREMS

We assume an elementary knowledge of logic and set theory. The following symbols will be used: "not" for negation, "and" for conjunction, "or" for disjunction, "$\rightarrow$" for implication, "$\leftrightarrow$" for equivalence, "$=$" for identity, "$(\exists X)$" for the existential quantifier, there exists an X, and "(X)" for the universal quantifier, for all X.

The system of axioms for set theory which we adopt is essentially the system Σ given in Gödel [1] [1] with the following exceptions:

I. We permit the existence of individuals.

II. Unless specifically indicated, we do not assume the axiom of regularity (Axiom D in Gödel). Theorems in which it is used in the proof will be indicated by a footnote.

Formally: the primitive notions are class, set, individual and the element relation ϵ. The convention is made that capital letters, X, Y, . . ., denote variables whose ranges consist of all classes and all individuals, and the lower case letters, x, y, . . ., denote variables whose ranges consist of all sets and all individuals.

The axioms of Group A in Gödel [1], are modified as follows:

1. If x is a set then x is a class.
2. If $X \epsilon Y$ then X, is a set or X is an individual.
3. The principle of extensionality:
   (a) If X and Y are classes then (u) $[u \epsilon X \leftrightarrow u \epsilon Y]$ implies $X = Y$.
   (b) If X is an individual then (Y) $[Y \notin X]$. (An individual does not contain any elements.)

[1] Gödel's axiom system is essentially the same as the system of Bernays [1]. Axiom D is due to von Neumann [2] (Axiom VI 4).

Because of 3(b) we make the convention that the range of a variable which immediately follows ϵ does not include individuals.

4. The existence of the unordered pair:

$$(x)(y)(\exists z)(u)(u \in z \leftrightarrow u = x \text{ or } u = y].$$

The remaining axioms are just as they appear in Gödel.

(Axiom D. The Axiom of Regularity: If X is non-empty then there is a $u \in X$ such that u and X have no elements in common.)

Definitions and Theorems

1. (a) X is a *proper class* if X is not a set or an individual. $((X)(Y)[Y \textit{ a proper class} \to Y \notin X])$.
   (b) If X and Y are classes, X is said to be a *subclass* of Y, $X \subseteq Y$, if every element of X is an element of Y. $((u)[u \in X \to u \in Y])$. X is said to be a *proper subclass* of Y, $X \subset Y$, if $X \subseteq Y$ and $X \neq Y$. (If X is a set it is called a *subset* of Y.)
   (c) The *power class* of X, $\mathscr{P}(X)$, is the class of all subsets of X. $(u \in \mathscr{P}(X) \leftrightarrow u \subseteq X.)$
   (d) The *empty set*, Λ, is defined by the properties that Λ is a set and $(X)[X \notin \Lambda]$.
   The *universe*, V, is defined by the property $(x)[x \in V]$.
   (e) We use the standard notation to denote classes. That is, $\{x, y, \ldots\}$ is the set whose elements are $x, y, \ldots$, and if $\mathscr{A}(x_1, x_2, \ldots, x_n)$ is uniquely defined for each $x_1, x_2, \ldots, x_n$ and P is a predicate then
   $u \in \{\mathscr{A}(x_1, x_2, \ldots, x_n) : P(x_1, x_2, \ldots, x_n)\} \leftrightarrow$
   $(\exists x_1)(\exists x_2)\ldots(\exists x_n)[u = \mathscr{A}(x_1, x_2, \ldots, x_n) \textit{ and } P(x_1, x_2, \ldots, x_n)]$.
   (f) A *unit set* is a set which has only one element. An *unordered pair* is a set which has two elements.
   (g) An *ordered pair*, $\langle x, y \rangle$ is defined as follows:
   $\langle x, y \rangle = \{\{x\}, \{x, y\}\}$.
   An *ordered n-tuple*, $\langle x_1, x_2, \ldots, x_n \rangle = \langle x_1, \langle x_2, \ldots, x_n \rangle \rangle$, where $n \geq 2$ is a positive integer.
   (h) The *Cartesian product*, $X \times Y = \{\langle x, y \rangle : x \in X \textit{ and } y \in Y\}$.
   (i) The *union*, $X \cup Y = \{u : u \in X \textit{ or } u \in Y\}$.
   (j) The *intersection*, $X \cap Y = \{u : u \in X \textit{ and } u \in Y\}$.

(k) The *difference*, $X \sim Y = \{u\colon u \in X \text{ and } u \notin Y\}$.

(l) A class X is said to be *complete* if for all $u \in X$, $u \subseteq X$.

(m) X and Y are said to be *disjoint* if $X \cap Y = \Lambda$. A class X is said to be *disjoint* if each pair of its elements is disjoint.

Because of the definitions of $\subseteq$, $\cup$, $\cap$ and $\times$ we make the convention that the range of a variable which immediately precedes or follows $\subseteq$, $\cup$, $\cap$ or $\times$ does not include individuals.

2. R is said to be a (*binary*) *relation* if R is a class of ordered pairs. R is said to be an *n-ary relation* if R is a class of ordered n-tuples. We use the symbol $x\,R\,y$ to mean the same as $\langle x, y\rangle \in R$.

In what follows R is meant to be a binary relation unless otherwise specified.

(a) The *domain* of R, $\mathscr{D}(R) = \{x\colon (\exists y)[\langle x, y\rangle \in R]\}$.
The *range* of R, $\mathscr{R}(R) = \{y\colon (\exists x)[\langle x, y\rangle \in R]\}$.

(b) $R''X = \{t\colon (\exists u)[u \in X \text{ and } \langle u, t\rangle \in R\}$.

(c) R is a *function* if whenever $\langle x, y\rangle \in R$ and $\langle x, z\rangle \in R$ then $y = z$.
R is a *function mapping* X *into* Y if R is a function and $\mathscr{D}(R) = X$ and $\mathscr{R}(R) \subseteq Y$.
R is a *function mapping* X *onto* Y if R is a function and $\mathscr{D}(R) = X$ and $\mathscr{R}(R) = Y$.

If R is a function often instead of $x\,R\,y$ or $\langle x, y\rangle \in R$ we write $R(x) = y$ or $R_x = y$.

(d) R is a 1-1 *function* if R is a function and whenever $\langle x, z\rangle \in R$ and $\langle y, z\rangle \in R$, $x = y$.

(e) X is *equivalent* to Y, $X \approx Y$, if there exists a 1-1 function mapping X onto Y.
$X \preccurlyeq Y$ if there exists a class $Y_0 \subseteq Y$ such that $X \approx Y_0$.
$X \prec Y$ if $X \preccurlyeq Y$ and not $X \approx Y$.

(f) If n is a natural number, $n > 1$, R is an *n-ary operation* on X if R is a function which maps $\{\langle x_1, x_2, \ldots, x_n\rangle\colon x_1, x_2, \ldots, x_n \in X\}$ into X. A 0-ary operation on X is a constant and a unary operation on X is a function which maps X into X. (We shall assume throughout the text that no ambiguity arises as to the order of an operation. Such ambiguities could be avoided, but the definition would be unnecessarily complicated)

(g) R is a *class-valued function* if whenever $x \in \mathscr{D}(R)$ and $\langle x, y\rangle \in R$, either $y = \Lambda$ or y is a unit set. We define

$$R[x] = \{y: \langle x, \{y\}\rangle \in R\}.$$

($R[x]$ can either be Λ, a set, or a proper class. We make this rather artificial definition so that if $R[x] = \Lambda$ we can distinguish the two possibilities that $x \notin \mathscr{D}(R)$ (then $R''\{x\} = \Lambda$) or if $\langle x, \Lambda\rangle \in R$ (then $R''\{x\} = \{\Lambda\}$)).

(h) If F is a function[1]:
union, $\bigcup_{x \in X} F(x) = \bigcup F''X = \{u: (\exists x)[x \in X \textit{ and } u \in F(x)]\}$,
intersection, $\bigcap_{x \in X} F(x) = \bigcap F''X = \{u: (x)[x \in X \rightarrow u \in F(x)]\}$,
Cartesian product, $\times_{x \in X} F(x) = \{f: f \textit{ is a function and } \mathscr{D}(f) = X \textit{ and } (x)[x \in X \rightarrow f(x) \in F(x)]\}$,
power, $X^Y = \{f: f \textit{ is a function and } \mathscr{D}(f) = Y \textit{ and } \mathscr{R}(f) \subseteq X\}$.

(Note. $X \times X \neq X^2$ but $X \times X \approx X^2$, and when n is finite, $X^n \approx \{u: u = \langle x_1, x_2, \ldots, x_n\rangle, \textit{ where } x_1, x_2, \ldots, x_n \in X\}$. Also it can be shown that $2^x \approx \mathscr{P}(x)$.)

(i) $R|X = \{\langle x, y\rangle: \langle x, y\rangle \in R \textit{ and } x \in X\}$.
(If R is an n-ary operation $R|X = \{\langle x, y\rangle : \langle x, y\rangle \in R, x = \langle x_1, x_2, \ldots, x_n\rangle, \textit{ and } x_1, x_2, \ldots, x_n \in X\}$. See parenthetical remark following 2(f).)

(j) *inverse*, $R^{-1} = \{\langle x, y\rangle: \langle y, x\rangle \in R\}$.

(k) *complement*, $\bar{R} = \{\langle x, y\rangle: \langle x, y\rangle \notin R\}$.

(l) *composite product*, $R \circ S = \{\langle x, y\rangle: (\exists z)[\langle x, z\rangle \in R \textit{ and } \langle z, y\rangle \in S\}$.

(m) the *identity relation*, $I = \{\langle x, y\rangle: x = y\}$.

(n) R is *reflexive on* X if $(x)[x \in X \rightarrow x\, R\, x]$.
R is *irreflexive on* X if $(x)[x \in X \rightarrow \textit{not } x\, R\, x]$.

[1] It is sometimes necessary to consider the union (intersection, Cartesian product) of classes. This definition can be generalized to this case as follows:
If F is a class-valued function:

$$\textit{union}, \bigcup_{x \in X} F[x] = \{u: (\exists x)[x \in X \text{ and } u \in F[x]\}.$$

Similarly for intersection and Cartesian product.

(o) R is *symmetric on* X if $(x)(y)[(x, y \in X \text{ and } x\, R\, y) \rightarrow y\, R\, x]$. R is *anti-symmetric on* X if $(x)(y)[(x, y \in X \text{ and } x\, R\, y \text{ and } y\, R\, x) \rightarrow x = y]$.

(p) R is *transitive on* X if $(x)(y)(z)[(x, y, z \in X \text{ and } x\, R\, y \text{ and } y\, R\, z) \rightarrow x\, R\, z]$.

(q) R is *connected on* X if $(x)(y)[x, y \in X \rightarrow (x\, R\, y \text{ or } y\, R\, x \text{ or } x = y)]$.

(r) R is a *partial ordering*[1] *on* X if R is transitive on X.

(s) R is a *linear ordering*[1] *on* X if R is a partial ordering on X and R is connected on X.

(t) x is said to be an *R-first* or *R-smallest* (*R-last* or *R-largest*) element of X if $x \in X$ and for all $y \in X$, $x \neq y \rightarrow x\, R\, y (y\, R\, x)$.

(u) x is said to be an *R-minimal* (*R-maximal*) element of X if $x \in X$ and for all $y \in X$, $y\, R\, x \rightarrow x\, R\, y$ $(x\, R\, y \rightarrow y\, R\, x)$.

(v) if $Y \subseteq X$, x is said to be an *R-lower bound* (*R-upper bound*) of Y if $x \in X$ and for all $y \in Y$, $x \neq y \rightarrow x\, R\, y$ $(y\, R\, x)$.

(w) if $Y \subseteq X$, x is said to be a *strict R-lower bound* (*strict R-upper bound*) of Y if x is an R-lower bound (R-upper bound) of Y and for all $y \in Y$, not $y\, R\, x$ (not $x\, R\, y$).

(x) if $Y \subseteq X$, x is said to be an *R-greatest lower bound* or *infimum* (*R-least upper bound* or *supremum*) of Y if x is an R-lower bound (R-upper bound) of Y and if y is an R-lower bound (R-upper bound) of Y and $x \neq y$ then $y\, R\, x$ $(x\, R\, y)$.

(y) R is said to be a *well-ordering on* X if R is antisymmetric on X, transitive on X, and every non-empty subclass of X has an R-first element [2]. (If every non-empty subclass of X has a unique R-first element then R is automatically anti-symmetric and transitive.)

(z) If R is a well-ordering on X and S is a well-ordering on Y then X is said to be *similar* to Y, $X \simeq Y$, if there exists a

[1] Some mathematicians require that a partial ordering be reflexive and/or anti-symmetric besides being transitive, but we do not make this restriction. The same remarks hold for a linear ordering.

[2] Some mathematicians require that a well-ordering be reflexive. Sometimes we shall need a reflexive well-ordering, sometimes an irreflexive well-ordering, and sometimes a well-ordering which may not be either reflexive or irreflexive.

1-1 function F mapping X onto Y such that for all u, $v \in X$, $u \neq v$, $u R v$ if and only if $F(u) S F(v)$.

(Note: A set X is said to be partially ordered, linearly ordered, well-ordered etc. if there exists a relation which is a partial ordering, linear ordering, well-ordering etc. on X.)

(z_1) A class X is *finite* if there exists a natural number n such that $X \approx n$. Otherwise X is infinite.

3. Ordinal numbers.
   (a) *X is an initial R-section of Y* if R is a relation, $X \subseteq Y$ and $Y \cap (R^{-1}{}''X) \subseteq X$. (That is, all R-predecessors in Y of members of X also belong to X.)
   (b) *X is the initial R-section of Y generated by u* if R is a relation, and $X = Y \cap R^{-1}{}''\{u\}$. (That is, X is the class of elements of Y which are R-predecessors of u. If R is transitive then an initial R-section of Y generated by u is an initial R-section of Y.)
   (c) X is an *ordinal number* [1] if
      (i) X is a set.
      (ii) X is irreflexively well-ordered by ϵ.
      (iii) For all $u \in X$, u is the initial ϵ-section of X generated by u.
   (d) *On* is the class of all ordinal numbers. $0 = \Lambda$.

Lower case greek letters, $\alpha, \beta, \gamma, \ldots$, will be used to denote variables whose range is *On*.
   (e) $\alpha < \beta$ if $\alpha \in \beta$; $\alpha \leq \beta$ if $\alpha < \beta$ or $\alpha = \beta$.
   (f) α is a *limit ordinal* if $\alpha = \bigcup_{\beta<\alpha} \beta$ and $\alpha \neq \Lambda$.

4. Properties of ordinal numbers:
   (a) If $\alpha \neq \Lambda$ then $\Lambda \in \alpha$ for all $\alpha \in On$.
   (b) Every element of an ordinal number is an ordinal number.
   (c) A set of ordinal numbers is an ordinal number if it is complete.
   (d) No ordinal number is an element of itself.
   (e) $\alpha < \beta$ if and only if $\alpha \subset \beta$.
   (f) For any two ordinal numbers α and β either $\alpha < \beta$, $\alpha = \beta$, or $\beta < \alpha$.

[1] The definition is that given by von Neumann [2] in 1928.

(g) *On* is well-ordered by ϵ.

(h) For every well-ordered set there is a unique ordinal number which is similar to it.

(i) If α is an ordinal number so is $\alpha \cup \{\alpha\}$.

(j) If α is not 0 or a limit ordinal then there exists a unique ordinal number β such that $\alpha = \beta \cup \{\beta\}$. We shall call β the *immediate predecessor* of α and designate it by $\alpha - 1$ and call α the *immediate successor* of β and designate it by $\beta + 1$.

(Note: All the above properties of ordinal numbers can be derived without using either the axiom of regularity or the axiom of choice.)

PART I

SET FORMS

1. The Well-Ordering Theorem

Before 1904, when Zermelo [1] published his proof that the axiom of choice implies the well-ordering theorem, the well-ordering theorem was considered as self-evident. Cantor and others used it without hesitation. WE 1–WE 3 are three classical forms of the well-ordering theorem. WE 4(m) and WE 5 are due to Azriel Levy [1], and WE 6 is a generalization of WE 5.

WE 1: *Every set can be well-ordered.*

WE 2: *Every set is equivalent to an ordinal number.*

WE 3: *Every set is equivalent to a subset of an ordinal number.*

Let m be a natural number, $m \geq 1$.

WE 4(m): *For every set x there exists an ordinal number α and a function f defined on α such that $\overline{\overline{f(\beta)}} \leqslant m$ for every $\beta < \alpha$ and* $\bigcup_{\beta<\alpha} f(\beta) = x$.

WE 5: *There exists a natural number $m \geq 1$ such that* WE 4(m).

WE 6: *For every set x there exists a natural number $m \geq 1$, an ordinal number α, and a function f defined on α such that $\overline{\overline{f(\beta)}} \leqslant m$ for every $\beta < \alpha$ and* $\bigcup_{\beta<\alpha} f(\beta) = x$.

WE 4(m) could be restated in the following way:
Every set is the union of a well-ordered set of finite sets each of which has not more than m elements. Similarly for WE 5 and WE 6.

[1] These forms are given in a paper entitled "Axioms of multiple choice" which has been submitted for publication to *Fundamenta Mathematicae*.

It is clear that WE 2 $\rightarrow$ WE 3 and WE 3 $\rightarrow$ WE 1. To prove that WE 1 $\rightarrow$ WE 2 see Gödel [1] [1]. The following equivalences are also clear: WE 1 $\leftrightarrow$ WE 4(1), WE 4(m) $\rightarrow$ WE 4(n) if $m \leq n$, and WE 4(m) $\rightarrow$ WE 5 $\rightarrow$ WE 6. Therefore, to prove that the above six forms are all equivalent it remains to prove the following:

THEOREM 1.1. WE 6 $\rightarrow$ WE 1 [2].

PROOF. Let x be an arbitrary set. For each set y, define a set of natural numbers N_y as follows:

(i) $m \in N_y$ if $(\exists f)(\exists \alpha)[f$ is a function, $\alpha \in On$, $\mathscr{D}(f) = \alpha$, $\bigcup_{\beta<\alpha} f(\beta) = y$, and $(\beta)[\beta < \alpha \rightarrow f(\beta) \preccurlyeq m]]$. (WE 6 implies that $N_y \neq \Lambda$ for every set y.)

Next, we shall prove the main result:

(ii) If $y \times y \subseteq y$ and m is a natural number such that $m > 1$ then $m \in N_y$ implies $m - 1 \in N_y$.

Suppose y and m satisfy the hypothesis of (ii) and $m \in N_y$. Then there exists a function f and an ordinal number α satisfying (i). Define

(iii) $u_{\beta\gamma\delta} = (f(\beta) \times f(\gamma)) \cap f(\delta)$, where $\beta, \gamma, \delta < \alpha$.

Now, $u_{\beta\gamma\delta}$ is a set of ordered pairs, so that it is a relation and we may talk about its domain and range. Moreover, we have

$$\mathscr{D}(u_{\beta\gamma\delta}) \subseteq f(\beta) \preccurlyeq m,\ \mathscr{R}(u_{\beta\gamma\delta}) \subseteq f(\gamma) \preccurlyeq m, \text{ and } u_{\beta\gamma\delta} \subseteq f(\delta) \preccurlyeq m.$$

Case 1. $(\beta)[\beta < \alpha$ and $f(\beta) \neq \Lambda \rightarrow (\exists\gamma)(\exists\delta)[\gamma, \delta < \alpha$, $\mathscr{D}(u_{\beta\gamma\delta}) \neq \Lambda$, and $\mathscr{D}(u_{\beta\gamma\delta}) \prec m]]$.

For each $\beta < \alpha$ with $f(\beta) \neq \Lambda$, let λ_β and μ_β be the lexicographically $<$ – first ordinal numbers γ and δ such that $\mathscr{D}(u_{\beta\gamma\delta}) \neq \Lambda$ and $\mathscr{D}(u_{\beta\gamma\delta}) \prec m$. (That is, first find ordinal numbers γ and δ which satisfy the conditions. Then let λ_β be the $<$ – smallest such γ which satisfies the conditions. Then given λ_β, let μ_β be the $<$ –

[1] The proof is given in Gödel [1] p. 27, theorem 7.7. This proof is quite similar to the proof of theorem 2.8 which we shall give in detail.

[2] The proof is due to Azriel Levy. See footnote p. 1.

smallest δ which satisfies the conditions.) Now define:

$$v_\beta = \begin{cases} \mathscr{D}(u_{\beta,\lambda_\beta,\mu_\beta}) & \text{if } f(\beta) \neq \Lambda, \\ \Lambda & \text{if } f(\beta) = \Lambda, \end{cases}$$

and $w_\beta = f(\beta) \sim v_\beta$. Next we define a function g as follows [1]:

$\mathscr{D}(g) = \alpha + \alpha$,

if $\beta < \alpha$ then $g(\beta) = v_\beta$,

if $\alpha \leq \beta$, and $\beta \sim \alpha \cong \gamma < \alpha$ then $g(\beta) = w_\gamma$.

We shall prove that $m - 1 \in N_y$ by proving that g is a function and $\alpha + \alpha$ an ordinal which satisfies (i). That is, we have to prove $\bigcup_{\beta<\alpha+\alpha} g(\beta) = y$ and if $\beta < \alpha + \alpha$ then $g(\beta) \preccurlyeq m - 1$. The first equality follows from set theoretic considerations and from the fact that $\bigcup_{\beta<\alpha} v_\beta \subseteq \bigcup_{\beta<\alpha} f(\beta) = y$. If $\beta < \alpha$, we have $v_\beta \prec m$ and consequently $g(\beta) \preccurlyeq m - 1$. If $\alpha \leq \beta < \alpha + \alpha$, then there is a $\gamma < \alpha$ such that $\beta \sim \alpha \cong \gamma$. If $f(\gamma) \neq \Lambda$, then $v_\gamma \neq \Lambda$, $f(\gamma) \preccurlyeq m$ by (i), and $v_\gamma \subseteq f(\gamma)$, therefore, $w_\gamma = f(\gamma) \sim v_\gamma \prec m$, so that $g(\beta) \preccurlyeq m - 1$. If $f(\gamma) = \Lambda$ then $w_\gamma = \Lambda$, so that again $g(\beta) \preccurlyeq m - 1$.

Case 2. $(\exists\beta)[\beta < \alpha,\ f(\beta) \neq \Lambda$ and $(\gamma)(\delta)[\gamma, \delta < \alpha$ and $\mathscr{D}(u_{\beta\gamma\delta}) \neq \Lambda \rightarrow \mathscr{D}(u_{\beta\gamma\delta}) \approx m]]$.

Let β be the $<$-smallest ordinal number in $\{\lambda\colon \lambda < \alpha,\ f(\lambda) \neq \Lambda$, and $(\gamma)(\delta)[\gamma, \delta < \alpha$ and $\mathscr{D}(u_{\lambda\gamma\delta}) \neq \Lambda \rightarrow \mathscr{D}(u_{\lambda\gamma\delta}) \approx m]\}$. Let $s \in f(\beta)$.

If $\gamma, \delta < \alpha$ and $\mathscr{D}(u_{\beta\gamma\delta}) \neq \Lambda$ then $\mathscr{D}(u_{\beta\gamma\delta})$ has m elements. It follows from (i) and (iii) that $u_{\beta\gamma\delta}$ has at most m elements ($u_{\beta\gamma\delta} \subseteq f(\delta) \preccurlyeq m$). Therefore $u_{\beta\gamma\delta} \approx m$ and $u_{\beta\gamma\delta}$ is a function.

Now, if in addition to $f(\beta) \neq \Lambda$, also $f(\gamma) \neq \Lambda$, then there exists a δ such that $u_{\beta\gamma\delta} \neq \Lambda$. (This follows from (iii) and the fact that $y \times y \subseteq y$.) Let δ_γ be the $<$-smallest ordinal number in $\{\delta\colon u_{\beta\gamma\delta} \neq \Lambda\}$.

Next, define

$$v_\gamma = \begin{cases} \{u_{\beta\gamma\delta}(s)\} & \text{if } f(\gamma) \neq \Lambda \\ \Lambda & \text{if } f(\gamma) = \Lambda \end{cases} \quad \text{and } w_\gamma = f(\gamma) \sim v_\gamma.$$

[1] The sum of two ordinal numbers is defined for example in Bachmann [1] p. 45.

(Note that since $f(\beta) \neq \Lambda$, if $\mathscr{D}(u_{\beta\gamma\delta}) \neq \Lambda$ then $s \in f(\beta) = \mathscr{D}(u_{\beta\gamma\delta})$ and $f(\gamma) \neq \Lambda$, therefore δ_γ exists.) Now define a function g on $\alpha + \alpha$ as follows:

If $\gamma < \alpha$ then $g(\gamma) = v_\gamma$.

If $\alpha \leq \gamma$ then $\gamma \sim \alpha \cong \lambda < \alpha$ and $g(\gamma) = w_\lambda$.

Again we shall show that g is a function and $\alpha + \alpha$ an ordinal number which satisfies (i) with m replaced by $m - 1$. As in case 1, we have $\bigcup_{\gamma < \alpha + \alpha} g(\gamma) = y$. If $\gamma < \alpha$, $g(\gamma) = v_\gamma \preccurlyeq 1 \preccurlyeq m - 1$. If $\alpha \leq \gamma$ and $\gamma \sim \alpha \cong \lambda$, $g(\gamma) = w_\lambda = f(\lambda) \sim v_\lambda$; if $f(\lambda) \neq \Lambda$, $v_\lambda \approx 1$ and $g(\gamma) \preccurlyeq m - 1$, while if $f(\lambda) = \Lambda$, $g(\gamma) = \Lambda \preccurlyeq m - 1$. Hence, if $\lambda < \alpha + \alpha$, then $g(\lambda) \preccurlyeq m - 1$. This completes the proof of (ii).

It follows from (ii) and mathematical induction that if $y \times y \subseteq y$ then y can be well-ordered. Now, to complete the proof of the theorem we must show that our original set x can be well-ordered. Therefore, we prove:

(iv) For every set x there exists a set y such that $x \cup (y \times y) \subseteq y$.

Construct y as follows:

Let $z_0 = x$ and $z_{n+1} = z_n \cup (z_n \times z_n)$. Let $y = \bigcup_{n=0}^{\infty} z_n$.

If $m < n$ then $z_m \subseteq z_n$ and $z_m \times z_m \subseteq z_n$. Clearly, $x \subseteq y$. Also,

$$\begin{aligned} y \times y &= \bigcup_{n,m=0}^{\infty} z_n \times z_m \\ &\subseteq \bigcup_{n,m=0}^{\infty} z_{\max(m,n)} \times z_{\max(m,n)} \\ &\subseteq \bigcup_{n,m=0}^{\infty} z_{\max(m,n)+1} \\ &\subseteq y. \end{aligned}$$

Hence, $x \cup (y \times y) \subseteq y$.

Therefore, x is a subset of a set which can be well-ordered, so that x can be well-ordered also, q. e. d.

2. The Axiom of Choice

Apparently, the first specific reference to the axiom of choice was given in 1890 in a paper of G. Peano [1]. In proving an existence theorem for ordinary differential equations, he ran across a situation in which such a statement is needed. In 1902 Beppo Levi [1], while discussing the statement that the union of a disjoint set t of non-empty sets has a cardinal number greater than or equal to the cardinal number of t [1], remarked that its proof depended on the possibility of selecting a single member from each element of t. Others, including Cantor, had used the principle earlier, but did not mention it specifically. We give here nine forms of the axiom of choice.

AC 1: *If s is a set of non-empty sets, there is a function f such that for every $x \in s$, $f(x) \in x$.*

AC 2: *If t is a disjoint set of non-empty sets, there is a set c which consists of one and only one element from each set in t.*

AC 3: *For every function f there is a function g such that for every x, if $x \in \mathscr{D}(f)$ and $f(x) \neq \Lambda$, then $g(x) \in f(x)$.*

AC 4: *For every relation r there is a function f such that $\mathscr{D}(f) = \mathscr{D}(r)$ and $f \subseteq r$.* (See P 12.)

AC 5: *For every function f there is a function g such that $\mathscr{D}(g) = \mathscr{R}(f)$ and for every $x \in \mathscr{D}(g)$, $f(g(x)) = x$.* (If f is 1-1 then g is its inverse.)

AC 6: *The Cartesian product of a set of non-empty sets is non-empty.*

Let m be a natural number, $m \geq 1$.

AC 7(m): *If s is a set of non-empty sets, there is a function f such that for every $x \in s$, $f(x) \neq \Lambda$, $f(x) \subseteq x$, and $\overline{f(x)} \leqslant m$.*

AC 8: *There exists a natural number $m \geq 1$ such that* AC 7(m).

AC 9: *If s is a set of non-empty sets, then there is natural number $m \geq 1$ and a function f such that for every $x \in s$, $f(x) \neq \Lambda$, $f(x) \subseteq x$, and $\overline{f(x)} \leqslant m$.*

In 1904, Zermelo [2] [1] stated a principle of choice similar to

[1] It is not known whether this statement is equivalent to the axiom of choice.

[2] Zermelo states in his paper that the proof of the theorem came out of conversations with Erhardt Schmidt.

AC 1 and proved that it implied the well-ordering theorem. (Zermelo took s to be the set of all non-empty subsets of a given set.) AC 3 and AC 6 are essentially restatements of AC 1.

In 1906, Bertrand Russell [1] gave a principle analogous to AC 2. He announced this principle as a possible substitute for Zermelo's, but he believed that it was weaker. (Russell actually stated his principle in the form of AC 6 with disjoint sets. However, by "product" Russell meant the following: $x \in \bigtimes_{a \in A} S_a$ if $x \subseteq \bigcup_{a \in A} S_a$ and $x \cap S_a$ is a unit set for each $a \in A$. This definition of product is due to Whitehead [1]. Hence Russell's form is the same as AC 2. See 2(h) in the section entitled Preliminary Definitions and Theorems for our definition of product.) In 1908, Zermelo, in [2], stated and in [3] proved that Russell's and his formulations of the axiom of choice are equivalent. The name "axiom of choice" is due to Zermelo [1], and the name "multiplicative axiom" or "multiplicative principle" is due to Russell [1]. AC 4 and AC 5 were given by Bernays [2] in 1941 [1]. AC 7(m) and AC 8 are due to Azriel Levy [2] and AC 9 is a generalization of AC 8.

We shall first show that AC 1–AC 6 are all equivalent. AC 1 obviously implies AC 2, and it is clear from the definition of product that AC 1 and AC 6 are equivalent.

THEOREM 2.1: AC 2 $\rightarrow$ AC 1.

PROOF. Let s be a set of non-empty sets. Define $t = \{\{x\} \times x : x \in s\}$. Then the choice set of AC 2 is the required choice function on s.

THEOREM 2.2: AC 1 $\rightarrow$ AC 3.

PROOF: Let f be an arbitrary function. Let $s = \mathscr{R}(f)$ and let F be the choice function on s. Define a function g so that for each $x \in \mathscr{D}(f)$, $g(x) = F(f(x))$. Then g is the required function.

THEOREM 2.3: AC 3 $\rightarrow$ AC 1.

[1] Bernays actually gives AC 4S and AC 5S which are stronger than AC 4 and AC 5. AC 4S is given as an axiom and AC 5S is proved equivalent to it.

[2] These forms are given in a paper entitled "Axioms of multiple choice" which has been submitted for publication to *Fundamenta Mathematicae*.

PROOF: Let s be a set of non-empty sets. Let f be a 1–1 function such that $\mathscr{R}(f) = s$. Define a function F such that for each $x \in s$, $F(x) = g(f^{-1}(x))$, where g is defined by AC 3. Then F is the required choice function.

THEOREM 2.4: AC 4 $\rightarrow$ AC 5.

PROOF: Let f be an arbitrary function and let $r = \{\langle x, y\rangle : \langle y, x\rangle \in f\}$. AC 4 implies that there is a function g such that $\mathscr{D}(g) = \mathscr{D}(r)$ and $g \subseteq r$. Clearly, for every $x \in \mathscr{R}(f) = \mathscr{D}(g)$, $f(g(x)) = x$.

THEOREM 2.5: AC 5 $\rightarrow$ AC 4.

PROOF: Let r be an arbitrary relation and define a function h as follows:

$$h = \{\langle\langle x, y\rangle, x\rangle : \langle x, y\rangle \in r\}.$$

AC 5 implies that there is a function g such that $\mathscr{D}(g) = \mathscr{R}(h)$ and for every $x \in \mathscr{D}(g)$, $h(g(x)) = x$. Now, $g(x)$ is an ordered pair, so we define $f(x)$ to be the second coordinate of $g(x)$, for each $x \in \mathscr{D}(g) = \mathscr{D}(r)$. Clearly, $\mathscr{D}(f) = \mathscr{D}(r)$, f is a function, and $f \subseteq r$.

THEOREM 2.6: AC 4 $\rightarrow$ AC 3.

PROOF: Let f be an arbitrary function. Define a relation r as follows:

$$r = \{\langle x, y\rangle : y \in f(x)\}.$$

AC 4 implies that there is a function g such that $\mathscr{D}(g) = \mathscr{D}(r)$ and $g \subseteq r$. g is the required function.

THEOREM 2.7: AC 3 $\rightarrow$ AC 4.

PROOF: Let r be an arbitrary relation. Define a function h as follows:

For each $x \in \mathscr{D}(r)$, $h(x) = \{y : \langle x, y\rangle \in r\}$. ($h(x) = r''\{x\}$.) AC 3 implies that there is a function f such that if $x \in \mathscr{D}(h)$ and $h(x) \neq \Lambda$, then $f(x) \in h(x)$. f is the required function.

Next, we shall show that each of AC 1–AC 9 is equivalent to a form of the well-ordering theorem. WE 1 implies AC 1, for WE 1 implies $\bigcup s$ can be well-ordered; then define the f of AC 1 to be the function which associates each set of s with its first element.

THEOREM 2.8 [1]: AC 1 $\to$ WE 2.

(The intuitive idea of the proof is as follows: Let x be a non-empty set and f a choice function on the set of all non-empty subsets of x. We well-order x as follows: Let $f(x) = a$ be the first element of x; $f(x \sim \{a\}) = b$ be the second element of x; $f(x \sim \{a, b\}) = c$ be the third element of x; etc.)

PROOF: Let x be a non-empty set. Let f be a choice function on the set of all non-empty subsets of x and define $f(\Lambda) = u$ where $u \notin x$. We define a function G as follows: For all ordinal numbers α, $G(\alpha) = f(x \sim G''\alpha)$. (It follows by transfinite induction that G is defined for all $\alpha \in On$.)

(1) G^{-1} is 1-1 on $\mathscr{R}(G) \cap x$. If $\alpha < \beta$, and $G(\alpha)$, $G(\beta) \in x$, then $G(\alpha) \neq G(\beta)$ since $G(\beta) = f(x \sim G''\beta) \in x \sim G''\beta$, and $G(\alpha) \in G''\beta$.

(2) There is an ordinal number α such that $G''\alpha = x$. For there must be some ordinal number β such that $G(\beta) \notin x$. If this were not so, by (1) G would be a 1-1 mapping of On into x which contradicts x being a set. Let α be the smallest ordinal number such that $\alpha \in \{\beta: G(\beta) \notin x\}$. Then by the definition of α, $G''\alpha \subseteq x$, and by the definition of G $x \sim G''\alpha = \Lambda$.

From (1) we see that an ordinal number satisfying (2) yields the desired result.

It is clear that AC 1 $\leftrightarrow$ AC 7(1), AC 7(m) $\to$ AC 7(n) if $m \leq n$, and AC 7(m) $\to$ AC 8 $\to$ AC 9. Therefore, it remains to be shown that AC 9 implies a form of the well-ordering theorem.

THEOREM 2.9: AC 9 $\to$ WE 6.

PROOF: (The proof is similar to the proof of 2.8.) Let x be a set and let s be the set of all non-empty subsets of x. By AC 9, there exists a natural number m and a function g such that for every $y \in s$, $g(y) \neq \Lambda$, $g(y) \subseteq y$, and $g(y) \preccurlyeq m$. Define $g(\Lambda) = u$ where $u \nsubseteq x$. Define a function G as follows: For all ordinal numbers α, $G(\alpha) = g(x \sim \bigcup_{\beta<\alpha} G(\beta))$.

Now, we proceed as in the proof of 2.8 and prove:

(1) G^{-1} is 1-1 on $\mathscr{R}(G) \cap \mathscr{P}(x)$ and

[1] The proof given here is a modification of Zermelo's [1] proof using ordinal numbers instead of well-ordered sets.

(2) There is an ordinal α such that $\bigcup G''\alpha = x$.
Then define $f = G|\alpha$, and we have m is the natural number, α is the ordinal number and f is the function which satisfies WE 6.

3. The Law of the Trichotomy

The equivalence of the axiom of choice and the trichotomy was given by Hartogs [1] in 1915. As in the case of the well-ordering theorem, the trichotomy was considered self-evident and was used without hesitation before 1915.

T: *For all sets x and y, either x is equivalent to a subset of y or y is equivalent to a subset of x.*

It is clear from the properties of ordinal numbers that WE 2 implies T. (See 4(f) on p. XXII.) To prove the implication the other way we first define *Hartogs' function* [1]:

DEFINITION 3.1: $\Gamma(x) = \{\alpha : \alpha \preccurlyeq x\}$.
We must prove that Γ is defined. That is,

LEMMA 3.2: *For all sets x, $\Gamma(x)$ is a set.*

PROOF: Let R be the set of all reflexive well-ordering relations on subsets of x. $R \subseteq \mathscr{P}(x \times x)$; therefore R is a set. The mapping which assigns to each well-ordering in R the corresponding ordinal number [2] is a mapping of R onto $\Gamma(x)$. Therefore $\Gamma(x)$ is a set.
We next prove

LEMMA 3.3: *For every set x, $\Gamma(x)$ is an ordinal number.*

PROOF: A set of ordinal numbers is an ordinal number if it is complete. That is. we must prove that if $\alpha \in \Gamma(x)$ then $\alpha \subseteq \Gamma(x)$.
Suppose $\alpha \in \Gamma(x)$ and $\beta \in \alpha$. Then $\alpha \preccurlyeq x$ and $\beta \subset \alpha$. Hence, $\beta \preccurlyeq x$ so that $\beta \in \Gamma(x)$.

THEOREM 3.4: T $\to$ WE 3.

PROOF: By T, either $x \preccurlyeq \Gamma(x)$ or $\Gamma(x) \preccurlyeq x$. The first inequality implies WE 3 since $\Gamma(x)$ is an ordinal number. The second is im-

[1] See Hartogs [1].
[2] See 2(*z*) and 4(*h*) in Preliminary Definitions and Theorems.

possible because it implies $\Gamma(x) \in \Gamma(x)$. (No ordinal number can be an element of itself.)

We shall next consider a proposition which is similar to T. It was stated by Lindenbaum in Lindenbaum and Tarski [1] and later proved by Sierpinski [5].

T′: *For every two non-empty sets, there is a mapping of one onto the other.*

Clearly T implies T′. Next we have the following

LEMMA 3.5: *For any two sets x and y, if there is a mapping of x onto y then there is a 1-1 function which maps $\mathscr{P}(y)$ into $\mathscr{P}(x)$.*

PROOF: Let f map x onto y. For $t \in \mathscr{P}(y)$, define $g(t) = f^{-1}{}''t$. Then g is a function from $\mathscr{P}(y)$ into $\mathscr{P}(x)$. To show that g is 1-1, we note that if $s, t \in \mathscr{P}(y)$, $g(s) \sim g(t) = f^{-1}{}''s \sim f^{-1}{}''t = f^{-1}{}''(s \sim t)$. Therefore, if $g(s) = g(t)$, $f^{-1}{}''(s \sim t) = \Lambda$ and $f^{-1}{}''(t \sim s) = \Lambda$; which implies $t \subseteq s$ and $s \subseteq t$; so that $s = t$.

THEOREM 3.6: T′ $\rightarrow$ WE 3.

PROOF: Consider the non-empty sets x and $\Gamma(\mathscr{P}(x))$. By T′, either there is a mapping of x onto $\Gamma(\mathscr{P}(x))$ or there is a mapping of $\Gamma(\mathscr{P}(x))$ onto x. If the first mapping exists, then 3.5 implies there is a 1-1 function which maps $\mathscr{P}(\Gamma(\mathscr{P}(x)))$ into $\mathscr{P}(x)$. But since $\Gamma(\mathscr{P}(x)) \prec \mathscr{P}(\Gamma(\mathscr{P}(x)))$, this implies that $\Gamma(\mathscr{P}(x)) \prec \mathscr{P}(x)$, which is impossible. Hence, there must be a mapping f of $\Gamma(\mathscr{P}(x))$ onto x. For each $t \in x$, define $g(t)$ to be the smallest element of $f^{-1}{}''t$. We see that g is a 1-1 function from x into $\Gamma(\mathscr{P}(x))$. Hence $x \preccurlyeq \Gamma(\mathscr{P}(x))$, which proves WE 3.

4. Maximal Principles

As mathematics developed further there also developed a need for another non-constructive proposition. A principle, which Kuratowski, Hausdorff, Zorn, and others, used to replace transfinite induction and the well-ordering theorem. It appears, at first glance, unrelated to the axiom of choice, but actually is equivalent to it.

This principle and principles similar to it are often referred to as forms of *Zorn's Lemma*. (It appears that Bourbaki [1] and Tukey [1] first referred to them in this way.) We prefer to call these forms *maximal principles* since it is a more descriptive title. Also, we discuss forms which are far afield from Zorn's original form, but still maximal principles.

In fact, it might be more appropriate to call the principle "Kuratowski's Lemma" or "Hausdorff's Lemma" rather than "Zorn's Lemma". In 1914, Hausdorff [1] derived M 5 (below) from the well-ordering theorem. M 6 is a specialization of M 5. In 1927, in the second edition of Hausdorff's book there is a derivation of a form similar to M 4 from the axiom of choice. Kuratowski [1] in 1922 derived M 4 and minimal principles equivalent to M 3 and M 4 from the well-ordering theorem [1]. (In our forms M 1–M 4 there is an obvious duality between minimal and maximal principles. We shall not discuss the minimal principles here.) Kuratowski used a minimal principle to prove a theorem in analysis. In 1932, R. L. Moore [1] derived a minimal principle equivalent to M 3 from the well-ordering theorem. (Moore lists Kuratowski in his bibliography.) It wasn't until 1935 that Zorn [1] published his paper. He was the first to state that a maximal principle implies the axiom of choice. He stated without proof that M 3 is equivalent to the axiom of choice (to be proven in another paper which was never published). Zorn was also the first one to apply a maximal principle in algebra. He discussed a list of applications of M 3. He was apparently unaware of the work of Hausdorff and Kuratowski.

O. Teichmüller [1] in 1939 and J. W. Tukey [1] in 1940 independently gave M 7 and proved it equivalent to the axiom of choice. Tukey also gave M 1, a generalization of M 3. Teichmüller was unaware of Zorn's work while Tukey refers to Zorn. N. Bourbaki [1] in 1939 gave a proposition similar to M 1 and states, without proof, that it is equivalent to the well-ordering theorem. (He stated a lemma similar to Kuratowski's lemma (see 4.11) to indicate how one would prove that M 1 implies the well-ordering theorem. M 7 is also stated by Bourbaki, but no proofs are given.) M 2 was given by Szele [1] in 1950.

There are two different forms of the maximal principles. M 1–M 4

[1] Some of the preliminary work in Kuratowski's paper is due to Hessenberg [1].

postulate the existence of a maximal element, and the remaining forms postulate the existence of a maximal set.

M 1: *If R is a transitive relation on a non-empty set x and if every subset of x which is linearly ordered by R has an R-upper bound, then there is an R-maximal element in x.*

M 2: *If R is a transitive relation on a non-empty set x and if every subset of x which is well-ordered by R has an R-upper bound, then there is an R-maximal element in x.*

DEFINITION 4.1: A *nest* is a class which is linearly ordered by inclusion, $\subseteq$.

M 3: *If every non-empty nest which is a subset of a non-empty set x has its union an element of x, then x has a maximal element*[1].

M 4: *If every well-ordered nest which is a subset of a non-empty set x has its union an element of x, then x has a maximal element*[1].

Clearly M 2 $\rightarrow$ M 1 $\rightarrow$ M 3 and M 2 $\rightarrow$ M 4 $\rightarrow$ M 3. There are many other maximal principles between M 2 and M 3 we could have chosen just by changing the ordering relation or the conditions on the upper bound. We chose M 1 since it is a common form and often used in practice and M 4 for historical reasons.

M 5: *If R is a transitive relation on x, then there exists a maximal subset*[1] *of x which is linearly ordered by R.*

M 6: *For every set x, there exists a maximal subset*[1] *of x which is a nest.*

DEFINITION 4.2: A non-empty property P is of *finite character* if a class X has the property P if and only if every finite subset of X has the property P. If X has the property P, we write $P[X]$.

(Note: If P is a property of finite character and $Q = \{x\colon x$ is finite and not $P[x]\}$ then $P[X]$ if and only if $x \in Q$ implies $x \nsubseteq X$. Conversely, if Q is any collection of non-empty finite sets and $P[X]$ if and only if $x \in Q$ implies $x \nsubseteq X$, then P is a property of finite character. The proofs of these statements follow directly from 4.2. In particular, the second statement implies that the property of an element not belonging to a set is a property of finite character.

[1] Maximal with respect to inclusion, $\subseteq$.

That is, if P is defined as follows: $P[X] \leftrightarrow u \notin X$ for some fixed element u, then P is a property of finite character. (To prove this take $Q = \{\{u\}\}$.))

M 7: *For every set x and every property P of finite character, there exists a maximal subset*[1] *of x which has the property P.*

Clearly M 7 → M 5 → M 6. In order to prove these seven forms are equivalent, it remains to be shown that M 3 → M 7 and M 6 → M 2. It turns out that it is easier to prove the latter implication in two steps, M 6 → M 5 and M 5 → M 2.

THEOREM 4.3: M 3 → M 7.

PROOF: Let x be a set and P be a property of finite character. Let $y = \{t: t \subseteq x \text{ and } P[t]\}$. We shall show that if $n \subseteq y$, $n \neq \Lambda$, and n is a nest then $\bigcup n \in y$. Clearly $\bigcup n \subseteq x$. Let u be a finite subset of $\bigcup n$. Then since n is a nest, u is a finite subset of some element of n. Since every element of n has the property P, it follows that $P[u]$. Therefore we obtain $P[\bigcup n]$ by the definition of finite character. Furthermore, the empty set has property P. Thus y satisfies the hypotheses of M 3, so y has a maximal element.

THEOREM 4.4: M 6 → M 5.

PROOF: Let R be a transitive relation on x. Let y be the set of all subsets of x which are linearly ordered by R. By M 6, there is a maximal subset n of y which is a nest. Let $m = \bigcup n$. We wish to show that m is a maximal element of y. Clearly $m \subseteq x$. Let u, $v \in m$, $u \neq v$; then $(\exists w)(u, v \in w \text{ and } w \in n)$. R linearly orders w, and hence either $u\,R\,v$ or $v\,R\,u$. Consequently R linearly orders m. Suppose m is not maximal; then for some $z \in y$, we have $m \subset z$. We observe that $n \cup \{z\}$ is a nest since every element of n is a subset of z. Since n is maximal, we must have $\{z\} \subseteq n$. That is, $z \in n$, whence $z \subseteq m$, which contradicts our assumption about z. Consequently, m is maximal, and therefore M 5 holds.

THEOREM 4.5: M 5 → M 2.

PROOF: Let R be a transitive relation on x. Let y be the set of all subsets of x which are well-ordered by R. Define a relation S on y as follows:

[1] Maximal with respect to inclusion, $\subseteq$.

$S = \{\langle t, u\rangle : t \in y$ and $u \in y$ and t is an initial R-section of $u\}$. S is clearly a transitive relation on y; therefore, by M 5 there is a maximal subset n of y which is linearly ordered by S. Let $m = \bigcup n$. We shall show that m has an R-upper bound, and that such an element is an R-maximal element of x.

First let us show that $m \in y$. Clearly $m \subseteq x$. Let z be any non-empty subset of m. Then for some $t \in n$, $z \cap t \neq \Lambda$. Let u be the R-first element of $z \cap t$. It remains to be shown that u is R-first in z. Now let $v \in z$, $v \neq u$. If $v \in t$, we have $u\, R\, v$ and not $v\, R\, u$ by the well-ordering property. If $v \notin t$, $v \in w$ for some $w \in n$. By the definition of S and the construction of n, t is an initial R-section of w. Hence $u\, R\, v$ and not $v\, R\, u$. Therefore, u is the unique R-first element of z so that R is a well-ordering on z and $m \in y$.

By an argument similar to that of the preceding theorem, m is maximal in y. Consequently, m cannot have a strict R-upper bound. For if b were a strict R-upper bound for m, we would have $m' = m \cup \{b\} \in y$ which contradicts the maximality of m. Since $m \in y$, it follows from the hypothesis of M 2 that m has an R-upper bound. Let b be such a bound. Suppose $z \in x$ and $b\, R\, z$. If not $z\, R\, b$, z would be a strict R-upper bound for m, which is impossible. Therefore, b is an R-maximal element of x, q. e. d.

There are some additional maximal principles which we shall discuss briefly. They are quite similar to the forms M 1–M 7 and first appeared in the literature in Kuratowski [1] in 1922. We shall call them M′ 1–M′ 7. The hypotheses for these forms are the same as that of the corresponding unprimed forms, but the conclusions are stronger. The conclusions for M′ 1–M′ 4 state that there exists a maximal element *larger than any given element*; the conclusions for M′ 5–M′ 7 state that there exists a maximal subset *containing any given subset* with the required property. We shall state only two of these additional forms in a formal way; it will be clear what the other forms are.

M′ 1: *If R is a transitive relation on x and if every subset of x which is linearly ordered by R has an R-upper bound and if $y \in x$ then there is an R-maximal element z in x such that either $y\, R\, z$ or $y = z$.*

M′ 7: *For every set x and every property P of finite character and for every subset y of x such that $P[y]$, there is a maximal subset*[1] *z of x such that $y \subseteq z$ and $P[z]$.*

It is clear that M′ $n \rightarrow$ M n for $n = 1, \ldots, 7$. Next we prove the implications the other way.

THEOREM 4.6: M $n \rightarrow$ M′n *for* $n = 1, 2, 3, 4$.

PROOF: Let $y \in x$ and let w be the set of all upper bounds for y. Then apply M n to the set w and thereby obtain M′n.

THEOREM 4.7: M 7 $\rightarrow$ M′n *for* $n = 5, 6, 7$.

PROOF: We give the proof for M′5. The others are analogous. Let y be a subset of x which is linearly ordered by R. "Linearly ordered" is a property of finite character. Call it P. We define a new property Q as follows. For any subset w of x, $Q[w]$ if and only $P[w \cup y]$. Q is clearly a property of finite character. By M 7 there is a maximal subset z of x which has the property Q. Therefore $z \cup y$ is a maximal subset of x which contains y and is linearly ordered by R.

The following maximal principals M 8 and M 10–M 13 are due to W. H. Gottschalk [1] and M 9 is due to A. D. Wallace [1]. (All these forms can be given in the primed form also.)

Let x be a set and R a relation. In the forms following, we shall omit these statements, as well as the statement that there is a maximal subset y of x satisfying the conclusion; just the conclusion will be stated in each form.

M 8: $y \times y \subseteq R$.

M 9: $y \times y \subseteq R \cup R^{-1}$.

M 10: $y \times y \subseteq \bar{R} \cup \bar{R}^{-1}$.

M 11: $y \times y \subseteq R \cup R^{-1} \cup I$.

M 12: $y \times y \subseteq \bar{R} \cup \bar{R}^{-1} \cup I$.

M 13: $y^n \subseteq R$. (By y^n here we mean $\{\langle t_1, t_2, \ldots, t_n\rangle : t_i \in y, i = 1, 2, \ldots, n\}$.)

(See 2(j), (k), and (m) in the section "Preliminary Definitions and Theorems" for the definition of R^{-1}, $\bar{R}$ and I.) Note that M 13

[1] Maximal with respect to inclusion, $\subseteq$.

refers to an n-ary relation, while in M 8–M 12, R is a binary relation.

M 5 could be stated in the following way: For every set x and transitive relation R, there is a maximal subset y of x such that $y \times y \subseteq R \cup R^{-1} \cup I$. In M 9, the set y is called coherent since for every s and t in y either $s\,R\,t$ or $t\,R\,s$. In M 10 y is called asymmetric since for every s and t in y neither $s\,R\,t$ nor $t\,R\,s$. In M 11, y is called chained or connected and in M 12, anti-symmetric.

It is clear that M 9 is equivalent to M 10 and M 11 is equivalent to M 12 for just substitute $\bar{R}$ for R. ($\bar{\bar{R}} = R$.) Moreover M 8 is a special case of M 13. Also M 8 $\rightarrow$ M 9 $\rightarrow$ M 11 and M 7 $\rightarrow$ M 13. The latter implication holds because if $P[y] \leftrightarrow y^n \subseteq R$ then P is a property of finite character. Finally, M 5 is a special case of M 11. We have now shown the following chain of implications holds:

M 7 $\rightarrow$ M 13 $\rightarrow$ M 8 $\rightarrow$ M 9 $\leftrightarrow$ M 10 $\rightarrow$ M 12 $\leftrightarrow$ M 11 $\rightarrow$ M 5.

We have previously shown that M 5 $\leftrightarrow$ M 7, (4.3, 4.4, 4.5, and the preceding discussion); hence M 1–M 13 are all equivalent.

We shall next show that these maximal principles are equivalent to the other forms of the axiom of choice we have given. Since there is so much in the literature about the equivalence of a maximal principle and the axiom of choice or the well-ordering theorem, we shall give several of these proofs. The first two (4.8 and 4.9) are well known.

Theorem 4.8: M 3 $\rightarrow$ AC 1.

Proof: Let S be a collection of non-empty sets, and let C be the collection of all choice functions on subsets of S. It is clear that every nest of functions in C has its union in C, and therefore, by M 3, C has a maximal element f. Clearly, f is the required choice function.

A similar proof could be given that M 3 $\rightarrow$ WE 1. Instead we shall prove

Theorem 4.9: M 1 $\rightarrow$ WE 1.

Proof: Let x be an arbitrary set, and let P be the set of all ordered pairs $\langle y, w\rangle$ such that $y \subseteq x$, $w \subseteq y \times y$ and w is a well-ordering on y. Let us define a relation R as follows: If $\langle y, w\rangle$ and $\langle y', w'\rangle$ are elements of P, then $\langle y, w\rangle R \langle y', w'\rangle$ if $y \subseteq y'$, $w = w' \cap (y \times y)$, and $y \times (y' \sim y) \subseteq w'$. That is, w' is an extension of w, $w = w'|y$, and y is an initial w'-section of y'. If s is an R-linearly ordered subset of P, then an R-upper bound for s is $\langle \bigcup \mathscr{D}(s), \bigcup \mathscr{R}(s)\rangle$. (We leave the proof that $\langle \bigcup \mathscr{D}(s), \bigcup \mathscr{R}(s)\rangle \in P$ and is an R-upper bound for s as an exercise.) Therefore, by M 1, P has an R-maximal element $\langle z, v\rangle$. Suppose $m \in x \sim z$. We define $z' = z \cup \{m\}$ and $v' = v \cup (z \times \{m\})$, and we observe that $\langle z, v\rangle R \langle z', v'\rangle$ and not $\langle z', v'\rangle R \langle z, v\rangle$. This contradicts the maximality of $\langle z, v\rangle$, so $z = x$ and hence v well-orders x, q. e. d.

We now proceed to the slightly more difficult task of proving that the axiom of choice implies a maximal principle. We shall first prove that the well-ordering theorem implies a maximal principle and then give several proofs that the axiom of choice implies a maximal principle.

THEOREM 4.10: WE 2 $\rightarrow$ M 5 [1].

The intuitive idea of the proof is as follows: Let x be a set and let R be a transitive relation on X. By WE 2, there is an ordinal number α such that α is equivalent to x. We construct a maximal R-linearly ordered subset of x by the following method: Let the first element of x belong. If the second element of x stands in the relation R to the first element, the second element is included, otherwise not. We continue by including those elements which are R-connected to the elements already included.

Formally, let F be a 1-1 function from α onto x. Define the function G as follows:

$G(\beta) = F(\beta)$ if $\beta < \alpha$ and $(\gamma)[\gamma < \beta \rightarrow (F(\beta)\, R\, G(\gamma)$ or $G(\gamma)\, R\, F(\beta))]$,
$G(\beta) = F(0)$ otherwise.

It is clear that the range of G is a maximal R-linearly ordered subset of x.

THEOREM 4.11: AC 3 $\rightarrow$ M′ 4 [2].

[1] This is the proof given in Hausdorff [1], 1914.

[2] The proof is due to Kuratowski [1], 1922. Similar proofs were given by Hausdorff [2] in 1927 and Bourbaki [2], Kneser [1], and Szele [1], all in 1950.

Let us first prove the following lemma, which is independent of the axiom of choice:

LEMMA: *Let $a \in x$ and suppose that every non-empty well-ordered nest of elements of x has its union in x. Furthermore, let f be a function on x to x, such that for all $y \in x$, $y \subseteq f(y)$. Then there is a $z \in x$ such that $a \subseteq z$ and $z = f(z)$.*

To prove this let f be a function on x to x such that for all $y \in x$, $y \subseteq f(y)$. Define a function G by transfinite induction as follows:

$$G(0) = a,$$

$$G(\alpha + 1) = f(G(\alpha)),$$

$$G(\alpha) = \bigcup_{\beta<\alpha} G(\beta) \text{ if } \alpha \text{ is a limit ordinal.}$$

We observe that if $\alpha < \beta$, $G(\alpha) \subseteq G(\beta)$. Since every non-empty well-ordered nest of elements of x has its union in x, G is a function from On into x. Since x is a set, the function G cannot be 1-1. Hence for some ordinals $\alpha < \beta$, $G(\alpha) = G(\beta)$. Therefore $G(\alpha) = G(\alpha + 1)$, and thus $G(\alpha) = f(G(\alpha))$. Since $G(0) = a$, $a \subseteq G(\alpha)$ and this completes the proof of the lemma.

Now let us use this lemma to prove the theorem. Let x be a non-empty set with the property that every well-ordered non-empty nest of elements of x has its union in x, and let $a \in x$. Define a function h on x as follows: For all $y \in x$, $h(y) = \{z\colon z \in x \text{ and } y \subset z\}$. By AC 3, there exists a function f such that $\mathscr{D}(f) = \mathscr{D}(h)$ and for all $y \in x$, if $h(y) \neq \Lambda$ then $f(y) \in h(y)$. If $h(y) = \Lambda$ define $f(y) = y$. Now, f satisfies the hypothesis of the lemma – f is a function from x to x such that for all $y \in x$, $y \subseteq f(y)$. Therefore, there is a $z \in x$ such that $a \subseteq z$ and $f(z) = z$. Such an element z is a maximal element of x.

THEOREM 4.12: AC 3 $\rightarrow$ M′ 5 [1].

PROOF: Let x be an arbitrary set, R a transitive relation on x, and y an R-linearly ordered subset of x. Let $L = \{z\colon z \subseteq x$ and R is a linear ordering on z and $y \subseteq z\}$. Suppose M′ 5 does not hold.

[1] The proof is due to Frink [1], 1952. This is similar to the proof given by Zermelo [2] in 1908.

Then there is no maximal element of L. By AC 3, there is a function f on L such that for each $z \in L$, $f(z) \in x \sim z$ and $z \cup \{f(z)\} \in L$. (Formally we obtain this function f as follows: For each $z \in L$ define $h_1(z) = \{w: w \in L \text{ and } z \subset w\}$. Because we are assuming that L has no maximal element, $h_1(z) \neq \Lambda$. AC 3 implies there exists a function h_2 such that $\mathscr{D}(h_1) = \mathscr{D}(h_2)$ and $h_2(z) \in h_1(z)$ for all $z \in L$. Now define $h_3(z) = h_2(z) \sim z$. Since $h_3(z) \neq \Lambda$ for all $z \in L$, again by AC 3, there exists a function f such that $\mathscr{D}(f) = \mathscr{D}(h_3) = L$ and for all $z \in L$, $f(z) \in h_3(z)$.) Let us denote $z \cup \{f(z)\}$ by z', which we shall call the *successor* of z.

Let $K \subseteq L$. We shall call K *complete* if

(a) $y \in K$,

(b) $z' \in K$ whenever $z \in K$,

(c) if N is a non-empty nest and $N \subseteq K$, then $\bigcup N \in K$.

Let $\mathscr{K}$ be the set of all complete sets. Since $L \in \mathscr{K}$, $\mathscr{K}$ is non-empty. Since the intersection of any non-empty set of complete sets is complete, $J = \bigcap \mathscr{K}$ is the smallest complete set. Let us now demonstrate that J is a nest.

If $z \in J$, z is *normal* if for every $u \in J$, $z \subseteq u$ or $u \subseteq z$. We shall prove that every element of J is normal, thereby proving that J is a nest.

Let z be a normal element of J. (J has a normal element since $y \in J$ is normal.) Define $K(z)$ to be the set of all $u \in J$ such that $u \subseteq z$ or $z' \subseteq u$. Clearly $y \in K(z)$ since $y \subseteq z$. Suppose $u \in K(z)$. If $z' \subseteq u$, then $z' \subseteq u'$ and $u' \in K(z)$. If $u \subseteq z$ and $u' \nsubseteq z$ then $z \subseteq u'$ since z is normal. But we have $u \subseteq z \subseteq u'$, which implies $u = z$ or $u' = z$. The latter equality is impossible since we have assumed $u' \nsubseteq z$. Consequently $u = z$, so $z' \subseteq u'$ and $u' \in K(z)$. It is also clear that if $N \subseteq K(z)$ is a nest then $\bigcup N \in K(z)$, because if there is an $n \in N$ such that $z' \subseteq n$ then $z' \subseteq \bigcup N$, otherwise $\bigcup N \subseteq z$. Therefore, $K(z)$ is a complete subset of J. Since J is the smallest complete set it follows that $K(z) = J$.

Now we are able to show that the set of normal elements form a complete set. First, y is normal. Next, suppose z is normal, then $J = K(z)$. Therefore, for all $u \in J$ either $u \subseteq z$, which implies $u \subseteq z'$, or $z' \subseteq u$. Consequently z' is normal. Finally, suppose

N is a nest of normal elements, (every subset of normal elements is a nest). Suppose $u \in J$. If every element of N is a subset of u then $\bigcup N \subseteq u$. If there is an $n \in N$ such that $u \subseteq n$ then $u \subseteq \bigcup N$. Hence, $\bigcup N$ is normal. Therefore, the set of normal elements of J forms a complete set which implies all elements of J are normal and thus J is a nest.

Since J is a nest and a complete class by (c), $\bigcup J \in J$, so that by (b) $(\bigcup J)' \in J$, which implies $(\bigcup J)' \subseteq \bigcup J$, which is a contradiction, q. e. d.

THEOREM 4.13: AC 1 $\rightarrow$ M 1 [1].

PROOF: Let x be an arbitrary non-empty set and R a transitive relation on x, which we may assume to be anti-symmetric and reflexive (since M 1 $\rightarrow$ M 1 (with R anti-symmetric and reflexive) $\rightarrow$ M3). For each R-linearly ordered subset y of x, let $\hat{y}$ be the set of all R-upper bounds of y in $x \sim y$. Let f be the choice function on the set of all non-empty sets of the type $\hat{y}$. If y is an R-linearly ordered subset of x, y is said to be an *f-chain* if $y \neq \Lambda$ and if for all $z \subset y$ with $\hat{z} \cap y \neq \Lambda$, $f(\hat{z})$ is an R-minimal element of $\hat{z} \cap y$. (At least one f-chain exists since $\{f(\hat{\Lambda})\}$ is an f-chain.)

The following properties hold for f-chains:

(a) If y is an f-chain, $z \subseteq y$, and $\hat{z} \cap y = \Lambda$ then $\hat{z} = \hat{y}$. (In fact all linearly ordered sets satisfy this condition.)

(b) If y is an f-chain and $\hat{y} \neq \Lambda$ then $y' = y \cup \{f(\hat{y})\}$ is also an f-chain.

To see this observe that if $z \subset y'$ and $\hat{z} \cap y' \neq \Lambda$ then either

(i) $z \subset y$ and $\hat{z} \cap y \neq \Lambda$

or (ii) $z \subseteq y$ and $\hat{z} \cap y = \Lambda$

or (iii) $z \nsubseteq y$.

Since y is an f-chain (i) implies that $f(\hat{z})$ is an R-minimal element of $\hat{z} \cap y'$. If (ii) holds then $\hat{z} \cap y' = \{f(\hat{y})\}$, but (a) implies $\hat{y} = \hat{z}$. Therefore, $f(\hat{z})$ is an R-minimal element of $\hat{z} \cap y'$. And (iii) is impossible because if $z \nsubseteq y$ and $z \subset y'$ then $f(\hat{y}) \in z$ which implies

[1] The proof is due to Weston [1], 1957.

$\hat{z} \cap y' = \Lambda$. Since (i), (ii), and (iii) are the only possibilities it follows that y' is an f-chain.

(c) For any two f-chains, one is an initial R-section of the other. Let y and z be f-chains and suppose $t \in z \sim y$. Let

$$u = \{s: s \in z \cap y \text{ and } s\, R\, t\}.$$

We shall show that $u = y$. Clearly $u \subseteq y$. Let us assume

(i) $u \subset y$.

It is also clear that $u \subseteq z$. Since $t \in z$ and $t \notin u$, we have

(ii) $u \subset z$.

Also, $t \in \hat{u} \cap z$, therefore

(iii) $\hat{u} \cap z \neq \Lambda$.

Since z is an f-chain, it follows from (ii) and (iii) that $f(\hat{u})$ is an R-minimal element of $\hat{u} \cap z$. In particular, since $t \in \hat{u} \cap z$,

(iv) $f(\hat{u})\, R\, t$.

Now, if $f(\hat{u}) \in y$ then since $f(\hat{u}) \in z$ and by (iv) we would have $f(\hat{u}) \in u$, which is impossible. Therefore,

(v) $f(\hat{u}) \notin y$.

If $\hat{u} \cap y \neq \Lambda$, since y is an f-chain, (i) implies $f(\hat{u}) \in \hat{u} \cap y \subseteq y$, which contradicts (v). Therefore,

(vi) $\hat{u} \cap y = \Lambda$.

By (i) and (vi) it follows from (a) that

(vii) $\hat{u} = \hat{y}$.

Now, suppose $r \in y \sim u$, then it follows from (vii) that there is an $s \in u$ such that $r\, R\, s$, (if not $r \in \hat{u}$ but $r \notin \hat{y}$, contradicting (vii)). Since $s \in u$, $s\, R\, t$, so that the transitivity of R implies $r\, R\, t$. Now, we have $r \in y$ and $r\, R\, t$, so that since $r \notin u$ we must have $r \in y \sim z$. Now, if we define a set

$$v = \{s: s \in y \cap z \text{ and } s\, R\, r\}$$

then by an argument analogous to the argument used to derive (vi), we obtain

(viii) $\hat{v} \cap z = \Lambda$.

But since $r\,R\,t$ and $t \in z$, $t \in \hat{v} \cap z$ which contradicts (viii). Therefore, $u = y$ and y is an initial R-section of z.

Since the set of all f-chains is a nest its union is linearly ordered by R. Let us next show that the union is an f-chain. Let $c = \bigcup\{y : y$ is an f-chain$\}$. Suppose $y \subset c$ and $\hat{y} \cap c \neq \Lambda$. Then there is an f-chain z such that $\hat{y} \cap z \neq \Lambda$ and an f-chain w such that $y \subseteq w$. We wish to show first that $y \subseteq z$. Since z and w are both f-chains it follows from (c) that either $w \subseteq z$ or z is an initial R-section of w. In the former case, $y \subseteq z$. In the latter case, if $y \nsubseteq z$ then there exists an element $s \in y$ such that $s \in w \sim z$ and $s \in \hat{z}$. This implies that for every $t \in \hat{y}$, $s\,R\,t$ which implies $\hat{y} \cap z = \Lambda$ which contradicts the choice of z. Hence, $y \subset z$. ($y \neq z$ since $\hat{y} \cap z \neq \Lambda$.)

Now, we have there exists an f-chain z such that $y \subset z$ and $\hat{y} \cap z \neq \Lambda$, therefore $f(\hat{y})$ is an R-minimal element of $\hat{y} \cap z \subseteq \hat{y} \cap c$. It remains to be shown that $f(\hat{y})$ is an R-minimal element of $\hat{y} \cap c$. Suppose not, then there is a $t \in \hat{y} \cap c$ such that $t \notin z$ and $t\,R\,f(\hat{y})$. Also, there is an f-chain w such that $t \in w$. Since z and w are both f-chains and since $t \in w \sim z$ it follows from (c) that z is an initial R-section of w, so that for every $s \in z$, $s\,R\,t$. This implies, in particular, $f(\hat{y})\,R\,t$. Consequently, since R is anti-symmetric, $t = f(\hat{y})$ which is a contradiction. Hence, $f(\hat{y})$ is an R-minimal element of $\hat{y} \cap c$ and c is an f-chain. Moreover, $\hat{c} = \Lambda$ because if $\hat{c} \neq \Lambda$ then we could construct $c' = c \cup \{f(\hat{c})\}$ which is also an f-chain by (b), and $c \subset c'$ which contradicts the choice of c.

Since c is linearly ordered by R the hypothesis of M 1 implies that c has an R-maximal element m. Clearly m is an R-maximal element of x, because if there exists an $n \neq m$ and $m\,R\,n$ then $n \in \hat{c}$ which is a contradiction, q. e. d.

Of the forms we have already given of the maximal principle, only two do not involve a relation or a property, and state that there is a maximal element or subset with respect to a natural relation among sets (M 3 and M 6). We shall concern ourselves only with the maximal subset form M 6. The question can then be raised as to whether the corresponding forms with other natural relations among sets are of equal strength. Let x and y be two arbitrary sets.

These sets are disjoint unions of the appropriate two of $x \cap y$, $x \sim y$, $y \sim x$. ($x = (x \sim y) \cup (x \cap y)$ and $y = (y \sim x) \cup (x \cap y)$). The immediate symmetric set-theoretical relations derivable from these are whether $x \cap y$ is the empty set, and how many of $x \sim y$ and $y \sim x$ are the empty set. If x and y are distinct and non-empty, it can readily be seen that exactly one of the relations $x \cap y = \Lambda$, one of $x \sim y$ and $y \sim x$ is empty ($x \subseteq y$ or $y \subseteq x$), and none of the three sets are empty ($x \nsubseteq y$ and $y \nsubseteq x$ and $x \cap y \neq \Lambda$), must hold. Accordingly we introduce the following notation:

Definition 4.14: Let X and Y be any two classes. Then

(i) $X\,D\,Y$ if $X \cap Y = \Lambda$.

(ii) $X\,\bar{D}\,Y$ if $X \cap Y \neq \Lambda$.

(iii) $X\,K\,Y$ if $X \subseteq Y$ or $Y \subseteq X$.

(iv) $X\,\bar{K}\,Y$ if $X \nsubseteq Y$ and $Y \nsubseteq X$.

(v) $X\,J\,Y$ if $X \nsubseteq Y$ and $Y \nsubseteq X$ and $X \cap Y \neq \Lambda$.

(vi) $X\,\bar{J}\,Y$ if $X \subseteq Y$ or $Y \subseteq X$ or $X \cap Y = \Lambda$.

Note that the properties $\bar{D}$, $\bar{K}$, and $\bar{J}$ are the complements of the properties D, K, and J respectively, and as we have previously observed, exactly one of D, K, and J must hold if the sets are non-empty, therefore, $\bar{D} = K$ or J, $\bar{J} = D$ or K, and $\bar{K} = D$ or J.

We shall also use the following consistent ambiguity: We shall say that a class has a given one of the above properties if every distinct pair of its elements has the property. If a class X has the property D, we shall write $D[X]$ and similarly for the others.

Let us now introduce the following notation:

M 14(A): *Every set has a maximal subset*[1] *which has the property A.*

G. Kurepa [1] has shown that M 14(A) is equivalent to the axiom of choice if A is any one of the properties $\bar{D}$, J or $\bar{J}$. M 14(D) was shown to be equivalent to the axiom of choice by R. Vaught [1]. M 14(K) is the same as M 6. But, J. D. Halpern has shown in 1961 in his Ph. D. thesis that M 14($\bar{K}$) is not equivalent to the axiom of choice. However, previously, Kurepa [1] had shown that the following proposition is equivalent to the axiom of choice:

[1] Maximal with respect to inclusion, $\subseteq$.

M 15($\bar{K}$): *Every set can be anti-symmetrically linearly ordered and every set has a maximal subset*[1] *which has the property* $\bar{K}$.

Clearly M 7 implies each of the forms M 14(A), $A = D, \bar{D}, K, J, \bar{J}$, and M 15($\bar{K}$), since the properties considered are all properties of finite character.

THEOREM 4.15: M 14($\bar{D}$) $\rightarrow$ M 14(D).

PROOF: Let x be an arbitrary set. For each element $s \in x$ we define a set N_s as follows:

$$N_s = \{u: u = \{s\} \text{ or } (\exists t)[t \in x,\ u = \{s, t\}, \text{ and } s\, D\, t]\}.$$

Now if $s \neq t$, then $N_s\, \bar{D}\, N_t$ if and only if $s\, D\, t$. M 14($\bar{D}$) implies that $\{N_s: s \in x\}$ has a maximal subset w which has the property $\bar{D}$. We also have that $N_s = N_t$ if and only if $s = t$; therefore, $\{s: N_s \in w\}$ is well-defined and is clearly a maximal subset of x which has the property D.

THEOREM 4.16: M 14(J) $\rightarrow$ M 14($\bar{D}$).

PROOF: Let x be an arbitrary set. For each $s \in x$, define $s_u = s \cup \{\langle s, u\rangle\}$, where u is a set such that $(t)(s)[(s \in x \text{ and } t \in x) \rightarrow \langle s, u\rangle \notin t]$. Therefore, if $s \neq t$ then $s_u\, \bar{K}\, t_u$, which implies $s_u\, J\, t_u \leftrightarrow s_u\, \bar{D}\, t_u$. If $s \neq t$, $s_u \cap t_u = s \cap t$, so that $s_u\, \bar{D}\, t_u \leftrightarrow s\, \bar{D}\, t$. M 14($J$) implies that $\{s_u: s \in x\}$ has a maximal subset which has the property $\bar{D}$.

THEOREM 4.17: M 14(D) $\rightarrow$ AC 2.

PROOF: Let s be a collection of non-empty disjoint sets. Let $x = \{\{\langle 0, v\rangle, \langle 1, w\rangle\}: w \in s \text{ and } v \in w\}$. The elements $\{\langle 0, v\rangle, \langle 1, w\rangle\}$ and $\{\langle 0, v'\rangle, \langle 1, w'\rangle\}$ of x are disjoint if and only if $w \neq w'$. By M 14(D), there exists a maximal subset y of x which has the property D. Then the choice set for the collection s can be taken to be $\{v: (\exists w)\{\langle 0, v\rangle, \langle 1, w\rangle\} \in y\}$, q. e. d.

THEOREM 4.18: M 14($\bar{J}$) $\rightarrow$ M 14(K).

PROOF: Let x be an arbitrary set and let $u \notin \bigcup x$. For each $s \in x$, define $s_u = s \cup \{u\}$. Then for all $s, t \in x$, $s_u \cap t_u \neq \Lambda$, so that $s_u\, \bar{J}\, t_u \leftrightarrow s_u\, K\, t_u$. Also $s_u\, K\, t_u \leftrightarrow s\, K\, t$ since $u \notin \bigcup x$. Therefore, if $\{s_u: s \in x\}$ has a maximal subset which has the property $\bar{J}$ then x has a maximal subset which has the property K, q. e. d.

[1] Maximal with respect to inclusion, $\subseteq$.

THEOREM 4.19: M 15($\bar{K}$) $\rightarrow$ AC 2.

PROOF: Let s be a family of non-empty disjoint sets. Let R be an anti-symmetric, linear ordering of $\bigcup s$. There is no loss of generality if we assume R is reflexive. If $x \in u \in s$, define

$$x_R = \{z\colon z \in u \text{ and } z\, R\, x\}.$$

Since R is anti-symmetric, if $x \neq y$ then $x_R \neq y_R$ for all $x, y \in \bigcup s$. Also

(a) $x_R\, \bar{K}\, y_R$ if and only if x and y belong to different elements of s.

Because if x and y belong to different elements of s then $x_R \cap y_R = \Lambda$ which implies $x_R\, \bar{K}\, y_R$. Conversely, suppose x and y belong to the same element of s. Since R is connected on $\bigcup s$ either $x\, R\, y$ or $y\, R\, x$. But $x\, R\, y$ implies $x \in y_R$ which implies $x_R \subseteq y_R$ and $y\, R\, x$ implies $y \in x_R$ which implies $y_R \subseteq x_R$. Therefore, if x and y belong to the same element of s then $x_R\, K\, y_R$.

Hence, it follows from (a) that a maximal subset of $\{x_R\colon x \in \bigcup s\}$ which has the property $\bar{K}$ yields a choice set for s, q. e. d.

We have now shown that the following implications hold:

M 7 $\rightarrow$ M 14(J) $\rightarrow$ M 14($\bar{D}$) $\rightarrow$ M 14(D) $\rightarrow$ AC 2, M 7 $\rightarrow$ M 14($\bar{J}$) $\rightarrow$ M 14(K) $=$ M 6, M 7 $\rightarrow$ M 15($\bar{K}$) $\rightarrow$ AC 2.

We have shown previously that M 6 $\rightarrow$ M 7 and AC 2 $\rightarrow$ M 7.

It has been shown by C. C. Chang [1] and Azriel Levy [1] that M 14(D) can be generalized by extending the property D to D_m, and, in an analogous way, we were able to generalize M 14($\bar{D}$).

DEFINITION 4.20: If m is a natural number, $m \geq 2$, a class X is said to have the property $D_m(\bar{D}_m)$ if every subset of X which has m distinct elements has an empty (non-empty) intersection. ($D_2 = D$ and $\bar{D}_2 = \bar{D}$.)

Now we claim that M 14(D_m) and M 14($\bar{D}_m$) are equivalent to the axiom of choice. Clearly M 7 $\rightarrow$ M 14($\bar{D}_m$) and the proof that

[1] Chang actually obtained a stronger form, M′ 14(D_m) but Levy in his unpublished paper (see footnote [1] on page 1), by using AC 7(m), was able to weaken it by eliminating the prime. We give here the weaker form, M 14(D_m).

M 14($\bar{D}_m$) $\to$ M 14(D_m) is similar to the proof that M 14($\bar{D}$) $\to$ M 14(D), 4.15.

THEOREM 4.21: M 14(D_m) $\to$ AC 7($m - 1$).

PROOF: Let x be a set of non-empty sets. We can assume without loss of generality that the elements of x are disjoint, since otherwise we can replace each element y of x by $\{\langle y, t\rangle : t \in y\}$.

Let

$$u = \{\{\langle 0, t\rangle, \langle 1, y\rangle\} : t \in y \text{ and } y \in x\}.$$

Let v be a maximal subset of u which has the property D_m. For each $y \in x$, define

$$f(y) = \{t : \{\langle 0, t\rangle, \langle 1, y\rangle\} \in v\}.$$

Then $f(y) \subseteq y$ since $v \subseteq u$; $f(y) \preccurlyeq m - 1$ since v has the property D_m; and $f(y) \neq \Lambda$ since $y \neq \Lambda$ and v is maximal, q. e. d. (See 4.17.)

Just as in the case of Levy's forms of the well-ordering theorem and the axiom of choice (WE 4(m) and AC 7(m)) we are able to obtain weaker forms of M 14(D_m).

M 16: *There exists a natural number* $m \geq 2$ *such that* M 14(D_m).

M 17: *For every set x there exists a natural number* $m \geq 2$ *such that x has a maximal subset*[1] *which has the property* D_m.

Clearly, we have M 14(D_m) $\to$ M 16 $\to$ M 17. Also, M 17 $\to$ AC 9; the proof is analogous to the proof of 4.21.

There is some difficulty involved in trying to obtain similar results for $\bar{D}_m$. Because of the definition of $\bar{D}_m$, (definition 4.20) the statement:

S: *For every set x, there exists a natural number* $m \geq 2$, *such that x has a maximal subset*[1] *which has the property* $\bar{D}_m$.

holds independently of the axiom of choice. We can prove S without making use of the axiom of choice. To see this let x be an arbitrary set.

Case 1. Every finite subset of x has a non-empty intersection. Then x itself is the required maximal set for every $m \geq 2$.

Case 2. There exists a finite subset y of x with n distinct elements which has an empty intersection. Then y is a maximal subset of x which has the property $\bar{D}_{n+1}$, so take $m = n + 1$.

[1] Maximal with respect to inclusion, $\subseteq$.

To eliminate this rather trivial case we modify the definition of $\bar{D}_m$ slightly as follows:

DEFINITION 4.22: If m is a natural number, $m \geq 2$, a class X is said to have the property $\bar{D}!_m$ if every subset of x which does not have more than m distinct elements has a non-empty intersection.

Therefore, if a class has the property $\bar{D}!_m$ every subset of m elements has a non-empty intersection, every subset of $m - 1$ elements has a non-empty intersection, etc. So that if $\bar{D}_m$ is replaced by $\bar{D}!_m$ in S, the proof of case 2 is not valid. In fact, we are able to prove that the following propositions are equivalent to the axiom of choice.

M 18: *There exists a natural number $m \geq 2$ such that* M 14($\bar{D}!_m$) [1].

M 19: *For every set x, there exists a natural number $m \geq 2$ such that x has a maximal subset* [2] *which has the property $\bar{D}!_m$.*

It is clear that M 7 → M 18 → M 19. We shall prove M 19 → M 17. The proof is a generalization of the proof of 4.15.

THEOREM 4.23: M 19 → M 17.

PROOF: Let us first observe that if every element of x is disjoint from every element of y, then a subset w of $x \cup y$ has the property D_m if and only if $w \cap x$ and $w \cap y$ both have the property D_m. Consequently, to prove that x has a maximal subset with the property D_m it is sufficient to prove the fact for

$$u = (\{\omega\} \times x) \cup \{\langle \alpha, \alpha \rangle : \alpha \in \omega\},$$

where ω is the set of natural numbers.

In what follows the range of the variables m, n, etc., is the set of natural numbers greater than 1.

The set u satisfies the following property:

(i) There is a 1-1 function φ from ω into u such that for each $s \in u$, there is an $\alpha \in \omega$ such that $s \cap \varphi(\alpha) = \Lambda$.

Note that (i) enables us to constructively enlarge any subset v

[1] It is rather easy to show that M 18 with $\bar{D}!_m$ replaced by $\bar{D}_m$ is also equivalent to the axiom of choice since, the proof that it implies M 16 is analogous to the proof of 4.15.

[2] Maximal with respect to inclusion, $\subseteq$.

of u with less than n elements to an n-element subset $\psi(n, v)$ with $\bigcap \psi(n, v) = \Lambda$.

Define

(ii) $z_n = \{t: t \subseteq u, t \approx n, \text{ and } \bigcap t = \Lambda\}$,

and for each $s \in u$, and n,

(iii) $f_n(s) = \{h: h \in \bigtimes_m z_m \text{ and } s \in h(n)\}$.

Let

(iv) $y = \bigcup_n f''_n u$.

·We observe that

(v) If $v \subseteq y$ then $\bigcap v \neq \Lambda$ if and only if for each n, (a) $v \cap f''_n u \prec n$ or (b) $v \cap f''_n u \approx n$ and $\bigcap f_n^{-1}{}''v = \Lambda$.

(v) is clearly true if $v = \Lambda$. Otherwise, if $h \in \bigtimes_{m=2}^{\infty} z_m$ then

$$\begin{aligned} h \in \bigcap v &\leftrightarrow (w)[w \in v \to h \in w] \\ &\leftrightarrow (n)(s)[s \in f_n^{-1}{}''v \to h \in f_n(s)] \\ &\leftrightarrow (n)(s)[s \in f_n^{-1}{}''v \to s \in h(n)] \\ &\leftrightarrow (n)[f_n^{-1}{}''v \subseteq h(n)]. \end{aligned}$$

Now $f_n^{-1}{}''v \approx v \cap f_n{}''u$ since f_n is 1-1 and if $n \neq m$ then $f_n{}''u \cap f_m{}''u = \Lambda$. By (ii) and (iii) $h(n) \approx n$ so that $f_n^{-1}{}''v \preccurlyeq n$. If $f_n^{-1}{}''v \approx n$ then $f_n^{-1}{}''v = h(n)$ and consequently has an empty intersection. Conversely, if for each n (a) or (b) is satisfied, we can define h by

(vi) $h(n) = \begin{cases} \psi(n, f_n^{-1}{}''v) & \text{if } f_n^{-1}{}''v \prec n \\ f_n^{-1}{}''v & \text{if } f_n^{-1}{}''v \approx n \end{cases}$

then $h \in \bigcap v$.

Now, let v be a maximal subset of y with the property $\bar{D}!_m$. This is equivalent to $v \cap f_n{}''u$ being a maximal subset of $f_n{}''u$ with the property $\bar{D}!_m$ for every n. Also, it follows from (i), that $v \cap f_m{}''u$ has at least m elements. But if $w \subseteq f_m{}''u$ has at least m elements, w has the property $\bar{D}!_m$ if and only if $f_m^{-1}{}''w$ has the property D_m. Thus, $f_m^{-1}{}''v$ is a maximal subset of u with the property D_m. The theorem follows easily.

Next, we turn to some more generalizations of the form **M** 14(A) and thereby answer some of the questions asked in Kurepa [1]. In order to simplify the notation we define a property K^* as follows:

DEFINITION 4.24: A class X is said to have the property K^* if X can be anti-symmetrically, linearly ordered and X has the property $\bar{K}$.

We shall show that M 14(D or K or $\bar{J}$), M 14(D or $\bar{J}$ or K^*), M 14($\bar{D}$ or J or $\bar{J}$ or K), and M 14($\bar{D}$ or J or K^*) are equivalent to the axiom of choice. Each of these forms also has a primed form which is equivalent to the axiom of choice, but we can obtain a weaker primed form.

M′ 20: *If x is an arbitrary set and $e \in x$ then there exists a maximal subset*[1] *y of x such that $e \in y$ and y has the property D or $\bar{D}$ or J or $\bar{J}$ or K or K^**[2].

Clearly M 14(D) implies M 14(D or K or $\bar{J}$), M 14(D or $\bar{J}$ or K^*), and M′ 20; and M 14($\bar{D}$) implies M 14($\bar{D}$ or J or $\bar{J}$ or K) and M 14($\bar{D}$ or J or K^*). We shall sketch the proofs that each of the above five forms imply the axiom of choice.

THEOREM 4.25: M 14(D *or* K *or* $\bar{J}$) $\to$ AC 2.

PROOF: Let S be an infinite set of disjoint, non-empty sets, no set of which is an element of another or of itself. (This latter property can be assumed without loss of generality for just replace each $t \in S$ by $t \times \{u\}$ where $u \notin \bigcup S$.)

Let C be the collection of all choice functions on subsets of S.

Let $T = \{\{s, t\} : t \in s \in S\}$.

Define $x_1 = \{u \times (a \cup f) : u \in T \text{ and } f \in C\}$ where a has the property that $a \cap \bigcup C = \Lambda$.

Now, apply M 14(D or K or $\bar{J}$) to the set x_1. Each of the alternatives imply the existence of the required choice set.

THEOREM 4.26: M 14(D *or* $\bar{J}$ *or* K^*) $\to$ AC 2.

PROOF: Let S and T be defined as in 4.25.

[1] Maximal with respect to inclusion, $\subseteq$.

[2] Strictly speaking M′ 20 is not a primed form since e is an element of x rather than a subset of x. So that the form we give is even weaker than the corresponding primed form.

Let P be the set of all reflexive linear orderings on subsets of $\bigcup S$.

Let $M = \{\{p\} \times s_t : p \in P$ and $[t \in s \in S$ and $s_t = \{r : r \in s$ and $r \, p \, t\}]\}$.

Define $x_2 = T \cup M$. (We may assume without loss of generality that T and M are disjoint.)

Apply M 14(D or $\bar{J}$ or K^*) to x_2.

THEOREM 4.27: M 14($\bar{D}$ *or* J *or* $\bar{J}$ *or* K) $\rightarrow$ AC 2.

PROOF: Let S, C, and T be defined as in 4.25.

Let $U = \{\{\{r\}\} \cup T_r : r \in T$ and $T_r = \{\{r, q\} : q \in T$ and $q \cap r = \Lambda\}\}$.

Define $x_3 = \{u \times (a \cup f) : u \in \mathrm{U}$ and $f \in C\}$, where a has the property that $a \cap \bigcup C = \Lambda$.

Apply M 14($\bar{D}$ or J or $\bar{J}$ or K) to x_3.

THEOREM 4.28: M 14($\bar{D}$ *or* J *or* K^*) $\rightarrow$ AC 2.

PROOF: Let S and T be defined as in 4.25, P and M as in 4.26 and U as in 4.27.

Let $W = \{u \times (b \cup m) : u \in U$ and $m \in M\}$, where b has the property that $b \cap \bigcup M = \Lambda$.

$$\text{Define } x_4 = \begin{cases} U & \text{if } M = \Lambda, \\ W & \text{if } M \neq \Lambda. \end{cases}$$

Apply M 14($\bar{D}$ or J or K^*) to x_4.

THEOREM 4.29: M′ 20 $\rightarrow$ AC 2.

PROOF: Let S be defined as in 4.25, x_2 as in 4.26, and x_3 as in 4.27.

Define $x = x_2 \cup x_3$. We may assume that x_2 and x_3 are disjoint.

Apply M′ 20 to x with $e \in x_3$.

The following propositions are put in here since they deal with the properties D, $\bar{D}$, J, $\bar{J}$, K and $\bar{K}$, and they were suggested by some problems proposed by Kurepa [1]. They are not the same type of maximal principles which were considered previously. Let us introduce the following schemata.

M 21($A : B$): *Any set which contains a maximal subset* [1] *with the property A contains a maximal subset* [1] *with the property B.*

[1] Maximal with respect to inclusion, $\subseteq$.

M 22$(A:B)$: *Every set can be anti-symmetrically, linearly ordered and any set which contains a maximal subset*[1] *with the property A contains a maximal subset*[1] *with the property B.*

Under the following conditions M 21$(A:B)$ is equivalent to the axiom of choice:

$A = D$ and $B = \bar{D}$, J, $\bar{J}$, or K; $A = \bar{D}$ and $B = D$, $\bar{J}$, or K; $A = J$ and $B = D$, $\bar{D}$, $\bar{J}$, or K; $A = \bar{J}$ and $B = \bar{D}$, J, or K; $A = K$ and $B = D$, $\bar{D}$, J, or $\bar{J}$; $A = \bar{K}$ and $B = D$, $\bar{D}$, J, $\bar{J}$ or K. It follows from J. D. Halpern's Ph. D. thesis that M 21$(A, \bar{K})$ is not equivalent to the axiom of choice if $A = D$, $\bar{D}$, J, $\bar{J}$, or K (See remarks following M 14(A)), and it is not known if M 21$(\bar{J}: D)$ or M 21$(\bar{D}: J)$ are equivalent to the axiom of choice.

However, M 22$(A:B)$ is equivalent to the axiom of choice for all A and B, $A \neq B$, such that $A = D$, $\bar{D}$, J, $\bar{J}$, K, or $\bar{K}$ and $B = D$, $\bar{D}$, J, $\bar{J}$, K, or $\bar{K}$.

Clearly, M 22$(A:B) \rightarrow$ M 21 $(A:B)$; M 14$(B) \rightarrow$ M 21$(A:B)$ and M 22$(A:B)$; and M 14$(\bar{K}) \rightarrow$ M 22$(A:\bar{K})$.

THEOREM 4.30: M 21$(D: K) \rightarrow$ M 14(K).

PROOF: Let x be an arbitrary set. Define

$$s_u = s \cup \{u\}, \text{ where } s \in x \text{ and } u \notin \bigcup x.$$

Let

$$S = \{s_u\colon s \in x\}.$$

If s_u, $t_u \in S$, $s_u \cap t_u = (s \cap t) \cup \{u\}$. Therefore, maximal sets in S which have the property D are one element sets. Hence, M 21$(D: K)$ implies that S contains a maximal subset T which has the property K. But if s_u, $t_u \in S$, $s_u\, K\, t_u \leftrightarrow s\, K\, t$, so that $\{s\colon s_u \in T\}$ is a maximal subset of x which has the property K. (See 4.18.)

THEOREM 4.31: M 21$(D: \bar{J}) \rightarrow$ M 14(K).

PROOF: Substitute $\bar{J}$ for K in 4.30. $\bar{J}$ and K are identical on S.

THEOREM 4.32: M 22$(D: \bar{K}) \rightarrow$ M 15$(\bar{K})$.

PROOF: Substitute $\bar{K}$ for K in 4.30.

THEOREM 4.33: M 21$(D: \bar{D}) \rightarrow$ M 14$(\bar{D})$.

[1] Maximal with respect to inclusion, $\subseteq$.

PROOF: Let x be an arbitrary set. (We shall assume that $\Lambda \notin x$.) Define

$s_u = s \cup \{\langle s, u\rangle\}$ where $s \in x$ and $(t)[(t \in x$ and $t \neq s) \rightarrow \langle s, u\rangle \notin t]$. $(s_u \cap t_u = s \cap t)$.

Let $v = x \times \{u\}$, $(s_u \cap v = \{\langle s, u\rangle\})$.

Define $S = \{s_u : s \in x\}$,

$$S_v = S \cup \{v\}.$$

If $s_u, t_u \in S$ then $s_u \bar{K} t_u$, therefore,

$$s_u\, J\, t_u \leftrightarrow s_u\, \bar{D}\, t_u \leftrightarrow s\, \bar{D}\, t, \text{ and } s_u\, \bar{D}\, v$$

$$(s_u\, \bar{J}\, t_u \leftrightarrow s_u\, D\, t_u \leftrightarrow s\, D\, t, \text{ and } s_u\, J\, v).$$

Now, we have $\{v\}$ is a maximal subset of S_v which has the property D. M 21$(D\colon \bar{D})$ implies that S_v contains a maximal subset T_v which has the property $\bar{D}$. $T_v = \{v\} \cup T$ where T is a maximal subset of S which has the property $\bar{D}$. Then $\{s\colon s_u \in T\}$ s a maximal subset of x which has the property $\bar{D}$. (See 4.16.)

THEOREM 4.34: M 21$(D\colon J) \rightarrow$ M 14$(\bar{D})$.

PROOF: Substitute J for $\bar{D}$ in 4.33. J and $\bar{D}$ are identical on S.

THEOREM 4.35: M 21$(\bar{D}\colon D) \rightarrow$ M 14(D).

PROOF: Let x be an arbitrary set. Define

$$x_u = x \cup \{u\} \text{ where } u \cap \bigcup x = \Lambda.$$

By definition of u, $\{u\}$ is a maximal subset of x_u which has the property $\bar{D}$. Therefore, M 21$(\bar{D}\colon D)$ implies that x_u contains a maximal subset y_u which has the property D. $y_u = y \cup \{u\}$ and y is a maximal subset of x which has the property D.

THEOREM 4.36: M 21$(\bar{D}\colon K) \rightarrow$ M 14(K).

PROOF: Substitute $\bar{D}$ for D in 4.30. Here, a maximal set in S which has the property $\bar{D}$ is S itself.

THEOREM 4.37: M 21$(\bar{D}\colon \bar{J}) \rightarrow$ M 14(K).

PROOF: Substitute $\bar{J}$ for K in 4.36. $\bar{J}$ and K are identical on S.

THEOREM 4.38: M 22$(\bar{D}\colon \bar{K}) \rightarrow$ M 15$(\bar{K})$.

PROOF: Substitute $\bar{K}$ for K in 4.36.

THEOREM 4.39: M 21$(\bar{D}: J) \rightarrow$ M 14$(\bar{K})$.

PROOF: Substitute J for $\bar{K}$ in 4.38. J and $\bar{K}$ are equivalent on S.

Theorem 4.39 implies that M 22$(\bar{D}: J) \rightarrow$ M 15$(\bar{K})$.

THEOREM 4.40: M 21$(K: \bar{D}) \rightarrow$ M 14$(\bar{D})$.

PROOF: Substitute K for D in 4.33, and ignore the v of 4.33. A maximal subset of S which has the property K is a one element set.

THEOREM 4.41: M 21$(K: J) \rightarrow$ M 14$(\bar{D})$.

PROOF: Substitute J for $\bar{D}$ in 4.40. J and $\bar{D}$ are identical on S.

THEOREM 4.42: M 21$(K: D) \rightarrow$ M 14(D).

PROOF: Substitute D for $\bar{D}$ in 4.40.

THEOREM 4.43: M 21$(K: \bar{J}) \rightarrow$ M 14(D).

PROOF: Substitute $\bar{J}$ for D in 4.42. $\bar{J}$ and D are identical on S.

THEOREM 4.44: M 22$(K: \bar{K}) \rightarrow$ AC 2.

PROOF: Let S be a family of non-empty disjoint sets. Since every set can be anti-symmetrically, linearly ordered, let R be a reflexive relation which anti-symmetrically, linearly orders $\bigcup S$. If $x \in u \in S$, define

$$x_u = \{z\colon z \in u \text{ and } z \mathrel{R} x\}.$$

Let

$$T = \{x_u\colon x \in u \in s\}.$$

If $x_u, y_u \in T$, $x \neq y \rightarrow x_u \neq y_u$ since $x \mathrel{R} y$ and $y \mathrel{R} x \rightarrow x = y$. Also, $x_u \mathrel{\bar{K}} y_v \leftrightarrow u \neq v$. Therefore, for each $u \in s$, $\{x_u\colon x \in u\}$ is a maximal subset of T which has the property K. M 22$(K: \bar{K})$ implies that T contains a maximal subset P which has the property $\bar{K}$. $\{x\colon x_u \in P\}$ is the required choice set. (See 4.19.)

THEOREM 4.45: M 21$(\bar{K}: \bar{D}) \rightarrow$ M 14$(\bar{D})$.

PROOF: Substitute $\bar{K}$ for D in 4.33 and ignore the u of 4.33. A maximal set in S which has the property $\bar{K}$ is S itself.

THEOREM 4.46: M 21$(\bar{K}: J) \rightarrow$ M 14$(\bar{D})$.

PROOF: Substitute J for $\bar{D}$ in 4.45. J and $\bar{D}$ are identical on S.

THEOREM 4.47: M 21$(\bar{K}: D) \to$ M 14 (D).

PROOF: Substitute D for $\bar{D}$ in 4.45.

THEOREM 4.48: M 21$(\bar{K}: \bar{J}) \to$ M 14(D).

PROOF: Substitute $\bar{J}$ for D in 4.47. $\bar{J}$ and D are identical on S.

THEOREM 4.49: M 21$(\bar{K}: K) \to$ M 14(K).

PROOF: Let x be an arbitrary set and let S be defined as in 4.30. Define

$$S^u = S \cup \{u\} \text{ where } u \notin \bigcup x.$$

$\{u\}$ is a maximal subset of S^u which has the property $\bar{K}$. M 21$(\bar{K}: K)$ implies that there exists a maximal subset T^u of S^u which has the property K. $T^u = T \cup \{u\}$. $\{s: s_u \in T\}$ is a maximal subset of x which has the property K.

THEOREM 4.50: M 21$(J: D) \to$ M 14(D).

PROOF: Substitute J for $\bar{D}$ in 4.35.

THEOREM 4.51: M 21$(J: K) \to$ M 14(K).

PROOF: Substitute J for $\bar{K}$ in 4.50. $\{u\}$ is also a maximal subset of S^u which has the property J.

THEOREM 4.52: M 21$(J: \bar{J}) \to$ M 14(K).

PROOF: Substitute $\bar{J}$ for K in 4.51. $\bar{J}$ and K are identical on S^u and S.

THEOREM 4.53: M 22$(J: \bar{K}) \to$ AC 2.

PROOF: Substitute J for K in 4.44. Maximal subsets of T with the property J are one element sets.

THEOREM 4.54: M 21$(J: \bar{D}) \to$ WE 1.

PROOF: Let x be an arbitrary set. Let $\mathscr{R}$ be the set of all reflexive well-orderings of subsets of x. If $R, R' \in \mathscr{R}$, R is said to be an extension of R' if $R' \subset R$ and $\mathscr{D}(R') \times (\mathscr{D}(R) \sim \mathscr{D}(R')) \subseteq R$. For each $R \in \mathscr{R}$, define

$$S_R = \text{set consisting of } R \text{ and all its extensions } (R = \bigcap S_R).$$

Let

$$\mathscr{S} = \{S_R: R \in \mathscr{R}\}.$$

Suppose S_R, $S_{R'} \in \mathscr{S}$ and $T \in S_R \cap S_{R'}$, then T is an extension of R and R', so that either R is an extension of R' and $S_R \subseteq S_{R'}$, or R' is an extension of R and $S_{R'} \subseteq S_R$. Hence the set $\mathscr{S}$ has the property that each pair of its elements either has the relation K or D. Therefore, maximal sets in $\mathscr{S}$ with the property J are one element sets. M 21$(J: \bar{D})$ implies that there exists a maximal subset $\mathscr{T}$ of $\mathscr{S}$ which has the property $\bar{D}$. Since $\bar{D}$ and K coincide on $\mathscr{S}$, this implies that $\mathscr{T}$ is a maximal subset of $\mathscr{S}$ which is a nest. Let

$$T = \bigcup_{S_R \in \mathscr{T}} R.$$

Since T is a union of comparable reflexive well-orderings of subsets of x it is clearly a well-ordering of a subset of x. Suppose $s \in x$ and $s \notin \mathscr{D}(T)$. Then $s \notin \mathscr{D}(R)$ for all $S_R \in \mathscr{T}$. Let

$$T^* = T \cup (\mathscr{D}(T) \times \{s\}) \cup \{\langle s, s\rangle\},$$

and let $\mathscr{T}^* = \mathscr{T} \cup \{T^*\}$. Since T is an extension of R for each $R \in \mathscr{T}$ and T^* is an extension of T, $\mathscr{T}^*$ is a nest. Since $T^* \notin \mathscr{T}$, this contradicts the maximality of $\mathscr{T}$. Therefore, $\mathscr{D}(T) = x$, and T is a reflexive well-ordering of x.

THEOREM 4.55: M 21$(\bar{J}: \bar{D}) \rightarrow$ WE 1.

PROOF: Substitute $\bar{J}$ for J in 4.54. A maximal subset of $\mathscr{S}$ with the property $\bar{J}$ is $\mathscr{S}$ itself.

THEOREM 4.56: M 21$(\bar{J}: K) \rightarrow$ WE 1.

PROOF: Substitute K for $\bar{D}$ in 4.55. K and $\bar{D}$ are identical on $\mathscr{S}$.

THEOREM 4.57: M 21$(\bar{J}: J) \rightarrow$ M 14$(\bar{D})$.

PROOF: Substitute $\bar{J}$ for D in 4.34. $\{v\}$ is a $\bar{J}$-maximal subset of S_v.

THEOREM 4.58: M 22$(\bar{J}: \bar{K}) \rightarrow$ AC 2.

PROOF: Substitute $\bar{J}$ for K in 4.44. A maximal set in T with the property $\bar{J}$ is T itself.

THEOREM 4.59: M 22$(\bar{J}: D) \rightarrow$ AC 2.

PROOF: Substitute D for $\bar{K}$ in 4.58. D and $\bar{K}$ are identical on T.

In most of the proofs of the theorems 4.30–4.59 sets were constructed in such a manner that either one element sets or the whole set had the desired maximal property. Then, in most of the cases the proofs were similar to previous cases, (4.9, 4.16, 4.18, and 4.19).

The final form to be considered in this section is an obvious extension of the idea of constructing maximal subsets with respect to "natural" relations. The relation to be considered is the ϵ-relation.

M 23: *Every set x contains a maximal subset*[1] *y such that for every s, t* ϵ *y, s* $\neq$ *t, either s* ϵ *t or t* ϵ *s.*

It is clear that M 7 $\rightarrow$ M 23 since M 23 deals with a property of finite character. We shall prove that M 23 $\rightarrow$ WE 1.

THEOREM 4.60: M 23 $\rightarrow$ WE 1.

PROOF: Let x be an arbitrary set and let $\mathscr{W} = \{W\colon (\exists y)[y \subseteq x$ and W is a reflexive well-ordering of $y]\}$. We define a relation R on $\mathscr{W}$ as follows:

If W, $W' \epsilon \mathscr{W}$ then

$$W\,R\,W' \leftrightarrow W \subset W' \text{ and } \mathscr{D}(W) \times (\mathscr{D}(W') \sim \mathscr{D}(W)) \subseteq W'.$$

(Note: W' is an extension of W, see 4.9 and 4.53.) For all $W \epsilon \mathscr{W}$ we define a function f on $\mathscr{W}$ as follows: If $W \epsilon \mathscr{W}$,

$$f(W) = \{\{W\}\} \cup \{f(W')\colon W'\,R\,W \text{ and } W' \epsilon \mathscr{W}\}.$$

We claim: (i) $f(W) = f(U) \leftrightarrow W = U$ and

(ii) $f(W) \epsilon f(U) \leftrightarrow W\,R\,U$.

It is clear that if $W = U$ then $f(W) = f(U)$. To prove the converse it is sufficient to prove that

$$\text{(iii)}\quad \{\{W\}\} \cap \{f(W')\colon W'\,R\,W \text{ and } W' \epsilon \mathscr{W}\} = \Lambda,$$

for all $W \epsilon \mathscr{W}'$. We leave the proof of (iii) as an exercise.

To prove (ii) we note that if $W\,R\,U$ then clearly $f(W) \epsilon f(U)$. Suppose $f(W) \epsilon f(U)$ then either $f(W) = \{U\}$ or $W\,R\,U$, but the first alternative is impossible.

[1] Maximal with respect to inclusion, $\subseteq$.

M 23 implies that $\mathscr{R}(f)$ contains a maximal subset Q such that for every S, $T \in Q$ either $S \in T$ or $T \in S$. Let $\mathscr{U} = \{U: f(U) \in Q\}$. Since f is a 1-1 mapping of $\mathscr{W}$ onto $\mathscr{R}(f)$ preserving order (by (i) and (ii)) it follows that $\mathscr{U}$ is a maximal subset of $\mathscr{W}$ such that for all U, $W \in \mathscr{U}$, either $U R W$ or $W R U$.

We claim that $\bigcup \mathscr{U}$ is a reflexive well-ordering on x. Since $\bigcup \mathscr{U}$ is a union of R-comparable reflexive well-orderings of subsets of x, $\bigcup \mathscr{U}$ is clearly a reflexive well-ordering on a subset of x. Suppose there exists an $s \in x$ such that $s \notin \mathscr{D}(\bigcup \mathscr{U})$. Then for all $U \in \mathscr{U}$, $s \notin \mathscr{D}(U)$. Let

$$V = \bigcup \mathscr{U} \cup (\mathscr{D}(\bigcup \mathscr{U} \times \{s\}) \cup \{\langle s, s\rangle\},$$

and let $\mathscr{U}^* = \mathscr{U} \cup \{V\}$, Since for each $U \in \mathscr{U}^*, U R V$, and $\mathscr{U} \subset \mathscr{U}^*$, the maximality of $\mathscr{U}$ is contradicted. Hence, $\bigcup \mathscr{U}$ is a reflexive well-ordering on x.

5. Algebraic forms

This section is devoted to algebraic forms which are equivalent to the axiom of choice. Many of these forms were used in abstract algebra long before it was known they were equivalent to the axiom of choice. However, it wasn't until 1953 that J. Schmidt [1] and D. Scott [1] independently proved that several of these forms were actually equivalent to the axiom of choice. We give eleven forms in this section. AL′ 2, AL′ 3, and AL′ 4 are given because of their generality and ease of application. The other forms refer to specific types of algebraic systems (lattices, Boolean algebras, etc.) or special types of properties (meet-irreducible subalgebras, absolutely dispensible elements, etc.).

DEFINITION 5.1: *Algebra.*

(1) $\mathscr{A} = \langle A, \{O_\beta: \beta \in K\}\rangle$ is called an *algebraic system* or an (*abstract*) *algebra* if A is a set and for each $\beta \in K$, O_β is a finitary operation defined on A. (A 0-ary operation is a constant.)

(2) $\mathscr{B} = \langle B, \{O_\beta|B: \beta \in K\}\rangle$ is called a *sub-algebra* of $\mathscr{A}$ if $B \subseteq A$ and B is closed with respect to O_β for all $\beta \in K$. (That is, if O_β is an n-ary operation then $O_\beta''\{\langle b_1, b_2, \ldots, b_n\rangle: b_i \in B, i = 1, 2, \ldots, n\} \subseteq B$.) $\mathscr{B}$ is called a *proper sub-algebra* if $\mathscr{B} \neq \mathscr{A}$.

When it will not lead to any confusion, we shall call the set "A" of $\mathscr{A}$ the algebra and the set "B" of $\mathscr{B}$ the sub-algebra.

DEFINITION 5.2: *Lattice.*

(1) $\mathscr{L} = \langle L, \{\vee, \wedge\}\rangle$ is called a *lattice* if the following conditions are satisfied for every a, b, and c in L [1]:

(a) L is a set and $\vee$ and $\wedge$ are binary operations,
(b) $a \vee b \in L$ and $a \wedge b \in L$,
(c) $a \vee a = a$ and $a \wedge a = a$,
(d) $a \vee b = b \vee a$ and $a \wedge b = b \wedge a$,
(e) $a \vee (b \vee c) = (a \vee b) \vee c$ and $a \wedge (b \wedge c) = (a \wedge b) \wedge c$,
(f) $a \vee (a \wedge b) = a$ and $a \wedge (a \vee b) = a$;

$\vee$ is called the *join* and $\wedge$ the *meet.*

(2) $1 \in L$ is called a *unit element* if for all $a \in L$, $a \wedge 1 = a$.

(3) $0 \in L$ is called a *zero element* if for all $a \in L$, $a \vee 0 = a$.

(4) $E \subseteq L$ is called an *ideal* if whenever $a, b \in E$, $a \vee b \in E$ and whenever $a \in L$ and $b \in E$, $a \wedge b \in E$. E is called a *proper ideal* if $E \neq L$. (If E is a proper ideal then $1 \notin E$.)

(5) If $a, b \in L$, $a \leq b$ if $a \vee b = b$ (it follows from 5.2, (1)(f) $a \vee b = b$ if and only if $a \wedge b = a$).

(6) $x = \bigvee_{a \in A} a$ if $a \leq x$ for all $a \in A$ and if $a \leq y$ for all $a \in A$ then $x \leq y$. (x is called the *least upper bound* of A.)

$x = \bigwedge_{a \in A} a$ if $x \leq a$ for all $a \in A$ and if $y \leq a$ for all $a \in A$ then $y \leq x$. (x is called the *greatest lower bound* of A.)

(7) The lattice $\mathscr{L}$ is called *complete* if for each non-empty subset A of L, $\bigvee_{a \in A} a \in L$ and $\bigwedge_{a \in A} a \in L$.

(8) $c \in L$ is called *compact* if whenever $c \leq \bigvee_{a \in A} a$ where $A \subseteq L$, then there exists a finite subset B of A such that $c \leq \bigvee_{a \in B} a$.

(Note: A lattice is an algebraic system, but as defined, a sub-

[1] This definition of a lattice is given in Birkhoff [1] p. 19.

algebra is not an ideal. However, if the lattice were defined as follows a subalgebra would be an ideal.

$\mathscr{L} = \langle L, \{\vee\} \cup \{O_a : a \in L\}\rangle$ where

O_a is a unary operation such that $O_a(b) = a \wedge b$, for all $a \in L$.)

AL 1: *Every lattice with a unit element and at least one other element contains a maximal proper ideal*[1] *as a subset.*

AL′ 1: *Every proper ideal of a lattice with a unit element can be extended to a maximal proper ideal*[1].

AL′ 2: *If A is an algebraic system, B a subalgebra and $a \in A$, but $a \notin B$, then there exists a maximal subalgebra*[1] *which contains B as a subset, but does not contain a as an element.*

DEFINITION 5.3: C is called a *finitary closure operator* if C is a monotone unary operation such that for every class X:

$$C(X) = \bigcup \{C(y) : y \subseteq X \text{ and } y \text{ is finite}\}.$$

If $C(X) \subseteq X$ then X is called *C-closed.*

AL′ 3: *If x is an arbitrary set, C a finitary closure operator, P a property of finite character, and y a C-closed subset of x which has the property P, then there is a maximal C-closed subset*[1] *of x which contains y as a subset and has the property P.*

AL′ 1 was given in 1953 by D. Scott [1] and AL′ 2 was given by Robert Blair in 1957 in an unpublished paper. AL′ 3 is a generalization of AL′ 1 and AL′ 2. AL′ 2 and AL′ 3 are false in the corresponding forms without primes as there may exist no subsets with the desired property.

We shall prove the following implications:

$$\text{M 3} \to \text{AL}' \text{ 3} \to \text{AL}' \text{ 2} \to \text{AL}' \text{ 1} \to \text{AL 1} \to \text{M 7}.$$

In section 4 it was shown that M 3 and M 7 are equivalent. The proof that M 3 → AL′3 is similar to the proof of 4.3. AL′ 1 is a special case of AL′ 2 and it is clear that AL′ 1 → AL 1.

THEOREM 5.4: AL′ 3 → AL′ 2.

[1] Maximal with respect to inclusion, $\subseteq$.

PROOF: Let $\mathscr{A} = \langle A, \{O_\beta : \beta \in K\}\rangle$ be an algebraic system. If $x \subseteq A$, define $C(x)$ as follows: $a \in C(x)$ if and only if there exist elements $a_1, a_2, \ldots, a_n \in x$ and an operation O_β, $\beta \in K$ such that $a = O_\beta\ (a_1, a_2, \ldots, a_n)$. Since O_β is a finitary operation for all $\beta \in K$, C is a finitary closure operator, and if B is a subalgebra of $\mathscr{A}$, then B is C-closed. Moreover, the property of excluding an element from a set is a property of finite character. Hence, AL′ 2 is a special case of AL′ 3.

THEOREM 5.5: AL 1 → M 7.

PROOF: Let x be a set and P a property of finite character. Let us assume that x does not have the property P, for if it does then it is the maximal subset which has the property P. Let us also assume that at least one non-empty subset of x has the property, otherwise the empty set is the required maximal set. Let $y = \{t : t \subseteq x \text{ and } P[t]\} \cup \{x\}$. If $s, t \in y$, define $s \wedge t = s \cap t$ and

$$s \vee t = \begin{cases} s \cup t & \text{if } P[s \cup t], \\ x & \text{otherwise,} \end{cases}$$

then $\langle y, \{\vee, \wedge\}\rangle$ is a lattice. The unit element of the lattice is x. By AL 1, $\langle y, \{\vee, \wedge\}\rangle$ has a maximal proper ideal z. We shall show that $\bigcup z$ is a maximal subset of x with the property P. Clearly, $\bigcup z \subseteq x$. Next, $x \notin z$ since z is a proper ideal. Let t be a finite subset of $\bigcup z$. Then there is a finite number of elements $s_1, s_2, \ldots, s_n$ of z such that $t \subseteq \bigcup_{i=1}^{n} s_i$. Since z is an ideal and $x \notin z$, $\bigcup_{i=1}^{n} s_i \in z$, which implies $P[\bigcup_{i=1}^{n} s_i]$. Consequently, $P[t]$, which implies $P[\bigcup z]$. $\bigcup z$ is maximal since z is maximal, q. e. d.

In the proof of 5.5 we just required AL 1 to hold for complete lattices.

AL′ 4 was given in 1959 by E. W. Beth [1]. It is one of the most general forms of the axiom of choice that we have come across. Before stating it, we give the following definition:

[1] AL′ 4 is a set theoretical interpretation of the syntactical form that Beth actually states.

DEFINITION 5.6:

(a) Let m and n be natural numbers and let R be an n-ary relation, then for $m < n$, $\langle x_1, x_2, \ldots, x_m \rangle \in \mathscr{D}_n^m(R)$ if there exist elements $x_{m+1}, x_{m+2}, \ldots, x_n$ such that $\langle x_1, x_2, \ldots, x_n \rangle \in R$.

(b) Let R be a quinary relation, then $R : X$ if for all natural numbers m and n, for all $u \in X^m$ and for all w such that $\langle w, m, n \rangle \in \mathscr{D}_5^3(R)$ there exists a $v \in X^n$ such that $\langle w, m, n, u, v \rangle \in R$. (Let $X^0 = \{\Lambda\}$ and $X^1 = \{0\} \times X$.)

AL′ 4: *If x is an arbitrary set, $y \subseteq x$, and $R : y$, then there is a maximal subset*[1] *z of x such that $y \subseteq z$ and $R : z$.*

The proof that M 3 → AL′ 4 is similar to the proof of 4.3. We shall show that AL′ 3 is a special case of AL′ 4.

THEOREM 5.7: AL′ 4 → AL′ 3.

PROOF: Let x be an arbitrary set, C a finitary closure operator, P a property of finite character, and y a C-closed subset of x which has the property P. Define a quinary relation R as follows:

$\langle w, m, n, u, v \rangle \in R \leftrightarrow m$ and n are natural numbers; $u \in x^m$; $v \in x^n$; and either $n = 0$, $w = \Lambda$, and $P[\mathscr{R}(u)]$, or $n = 1$, $w \in C(\mathscr{R}(u))$, and $v = \langle 0, w \rangle$.

Then, we claim, $R : z$ if and only if $P[z]$ and z is C-closed. (We leave the details of the proof to the reader.) Now, it is clear that AL′ 4 → AL′ 3.

The following maximal principle is due to Mrowka[2] [1] and deals with a special type of lattice, a Boolean algebra (see 5.2).

DEFINITION 5.8: *Boolean Algebra.*

$\mathscr{B} = \langle B, \{\vee, \wedge\} \rangle$ is said to be a *Boolean algebra* if the following conditions are satisfied:

(a) $\mathscr{B}$ is a lattice.

(b) $a \wedge (b \vee c) = (a \wedge b) \vee (a \wedge c)$ and $a \vee (b \wedge c) = (a \vee b) \wedge (a \vee c)$ for all $a, b, c \in B$.

(c) 0 and 1 are in B.

[1] Maximal with respect to inclusion, $\subseteq$.

[2] Mrowka showed that the axiom of choice implies AL 5, but his proof of the converse was incorrect. (This was corrected in Mrowka [2].)

(d) For every $a \in B$ there exists an element $b \in B$ such that $a \vee b = 1$ and $a \wedge b = 0$. (It can be shown that b is unique. b is called the *complement* of a and is denoted by $\sim a$.)

AL 5: *If B is a Boolean algebra and $S \subseteq B$ such that $0 \notin S$, then there exists a maximal proper ideal*[1] *disjoint from S.*

(See 5.2(4) for the definition of a proper ideal. Every non-empty ideal E in a Boolean algebra B contains 0, for if $a \in E$ then $\sim a \wedge a = 0 \in E$.)

The proof that AL′ 3 → AL 5 is analogous to the proof of 5.4 since a Boolean algebra can be defined as an algebraic system in such a way that an ideal is a subalgebra (see the remarks following 5.2).

THEOREM 5.9: AL 5 → M 7.

PROOF: Let x be an arbitrary set and P a property of finite character. Let $\mathscr{B} = \langle B, \{\vee, \wedge\}\rangle$ be defined as follows: $B = \mathscr{P}(x)$, $\vee = \cup$, and $\wedge = \cap$, $(0 = \Lambda$ and $1 = x)$, then $\mathscr{B}$ is the Boolean algebra of the set of all subsets of x. Let $S = \{y: y \subseteq x$ and not $P[y]\}$. Then $S \subseteq B$ and $\Lambda \notin S$ since $P[\Lambda]$ for every property of finite character P. Therefore, AL 5 implies that there exists a maximal proper ideal E disjoint from S. It is clear that $\bigcup E$ is the desired maximal subset of x.

We shall consider next six more algebraic equivalents of the axiom of choice.

AL 6: *If $\mathscr{L} = \langle L, \{\vee, \wedge\}\rangle$ is a complete lattice, $a, b, c \in L$, $a \leq b$, c is compact, $c \leq b$ and $c \nleq a$ then there exists an element $m \in L$ such that $a \leq m \leq b$, $c \nleq m$ and if there is a $d \in L$ such that $m < d \leq b$ then $c \leq d$* (in other words, m is a maximal element in the interval between a and b which is not greater than or equal to c).

Before stating the remaining forms in this section we require some further definitions.

DEFINITION 5.10: Let $\mathscr{A} = \langle A, \{O_\beta: \beta \in K\}\rangle$ be an algebraic system (see 5.1).

[1] Maximal with respect to inclusion, $\subseteq$.

(1) A subalgebra B is said to be *meet-irreducible* if whenever $B = \bigcap \mathscr{S}$ where $\mathscr{S}$ is a set of subalgebras then $B \in \mathscr{S}$.

(2) $B \subseteq A$ is said to be a *basis* for A if A is the smallest subalgebra containing B as a subset and whenever $C \subset B$ there exists a subalgebra $D \subset A$ such that $C \subseteq D$.

(3) An element $x \in A$ is said to be *absolutely dispensible* if for all sets $s \subseteq A$ whenever A is the smallest subalgebra containing S as a subset then A is the smallest subalgebra containing $S \sim \{x\}$ as a subset.

AL 7–AL 11 all refer to an algebraic system $\mathscr{A}$ and subalgebras in that system.

AL 7: *Every subalgebra is the intersection of all meet-irreducible subalgebras containing it as a subset.*

AL 8: *Every proper subalgebra has a proper meet-irreducible subalgebra containing it as a subset.*

AL 9: *The set of absolutely dispensible elements is equal to the intersection* [1] *of the set of all maximal proper subalgebras* [2].

AL 10: *For every element in a basis there exists a maximal subalgebra* [2] *which does not contain it as an element.*

AL 11: *If A has a finite basis and $B \subset A$ is a subalgebra then there exists a maximal proper subalgebra* [2] *which contains B as a subset.*

Diener [1], in 1956, proved AL 6 and AL 8 equivalent to a form of the axiom of choice. AL 7 was derived from the axiom of choice by Birkhoff and Frink [1] in 1948. However, there is an error in their proof (the proof of Lemma 4, pp. 301–2, is incorrect) which was later noted and corrected by Diener [1]. In 1938, McCoy [1] proved a special case of AL 7 and in 1949 Fuchs [1] [3] gave a correct proof of AL 7. The proof given here is similar to Diener's. AL 9 is due to Frattini [1] and Neumann [1]. Frattini in 1885 only proved the theorem for finite groups and did not require the axiom

[1] We make the convention here that $\bigcap A$ is the whole algebra.

[2] Maximal with respect to inclusion, $\subseteq$.

[3] Fuchs proved a special case, but his proof extends to the general case virtually unchanged.

of choice. The proof given here is essentially the same as the proof given by Neumann in 1937. Krull [1], in 1929, derived a proposition which is similar to AL 11 from the well-ordering theorem. In 1953 Schmidt [1] proved that AL 7, AL 9, AL 10, and AL 11 imply a form of the axiom of choice.

We shall prove

$$\text{M 3} \to \text{AL 6} \to \text{AL 7} \to \text{AL 8} \to \text{AC 3}$$

$$\text{AL}' \, 3 \to \text{AL 9} \to \text{AL 10} \to \text{M 7}$$

$$\text{AL}' \, 3 \to \text{AL 11} \to \text{M 7}.$$

It is clear that AL 7 $\to$ AL 8 and since an element in the basis is not absolutely dispensible, AL 9 $\to$ AL 10.

THEOREM 5.11: M 3 $\to$ AL 6.

PROOF: Let $\mathscr{L} = \langle L, \{\wedge, \vee\}\rangle$ be a complete lattice and for $a, b \in L$, $a \leq b$, let $[a, b] = \{d\colon d \in L \text{ and } a \leq d \leq b\}$. Suppose $c \in L$ is compact, $c \leq b$ and $c \nleq a$. Let $\mathscr{D} = \{D\colon D \subseteq [a, b] \text{ and } D \neq \Lambda$ and $c \nleq \bigvee D\}$. $\mathscr{D} \neq \Lambda$ since $\{a\} \in \mathscr{D}$. Let $\mathscr{N}$ be a non-empty nest contained as a subset in $\mathscr{D}$. We shall show that $\bigcup \mathscr{N} \in \mathscr{D}$. Clearly, if $\mathscr{N} \neq \Lambda$, then $\bigcup \mathscr{N} \neq \Lambda$ and $\bigcup \mathscr{N} \subseteq [a, b]$. Suppose $c \leq \bigvee \bigcup \mathscr{N}$, then since c is compact there exist $d_1, d_2, \ldots, d_k, N_1, N_2, \ldots, N_k$, $d_i \in N_i \in \mathscr{N}$ such that $c \leq \bigvee \{d_1, d_2, \ldots, d_k\}$. Since $\mathscr{N}$ is a nest, for some $j \in \{1, 2, \ldots, k\}$, $N_i \subseteq N_j$ for all $i = 1, 2, \ldots, k$, therefore $c \subseteq \bigvee N_j$. This is a contradiction since $N_j \in \mathscr{D}$. Hence $\bigcup \mathscr{N} \in \mathscr{D}$ so that by M 3, $\mathscr{D}$ has a maximal element M.

Let $m = \bigvee M$, then $a \leq m \leq b$ and $c \nleq m$. Suppose there is a $d \in [a, b]$ so that $m < d \leq b$. This implies $M \subset M \cup \{d\}$ so that $M \cup \{d\} \notin \mathscr{D}$. Therefore, $c \leq \bigvee (M \cup \{d\}) = d$, q. e. d.

Before proceding with the proof that AL 6 $\to$ AL 7 we prove a lemma due to Buchi [1].

DEFINITION 5.12: $\mathscr{L} = \langle L, \{\wedge, \vee\}\rangle$ is said to be an *algebraic lattice* if L is the set of all subalgebras of an algebraic system. The meet of two elements of L is ordinary set intersection, but the join of two elements is the smallest subalgebra containing them both. (This is usually not ordinary set union.)

It is easily verified that an algebraic lattice is a complete lattice.

LEMMA 5.13: *Every element in an algebraic lattice is the join of compact elements.*

PROOF: Let $\mathscr{L} = \langle L, \{\wedge, \vee\}\rangle$ be an algebraic lattice and let $\mu \in a \in L$. We wish to show that the subalgebra generated by μ, (the smallest subalgebra containing μ), denoted by (μ), is a compact element of L. Since (μ) is a subalgebra and L is the set of all subalgebras of an algebra, $(\mu) \in L$. Suppose $M \subseteq L$ and $(\mu) \leq \bigvee_{a \in M} a$, then $\mu \in \bigvee_{a \in M} a$. This implies that there exists a finite subset N of M such that $\mu \in \bigvee_{a \in N} a$. Since $\bigvee_{a \in N} a$ is a subalgebra it follows that $(\mu) \leq \bigvee_{a \in N} a$. Hence (μ) is compact.

Let $a \in L$, then clearly $a = \bigvee_{\mu \in a} (\mu)$, q. e. d.

THEOREM 5.14: AL 6 $\rightarrow$ AL 7.

PROOF: Let A be an algebraic system and $\langle L, \{\wedge, \vee\}\rangle$ the algebraic lattice of the set of all subalgebras of A. Let $B \in L$ and let D be the intersection of all meet-irreducible subalgebras containing B. Suppose $B < D$. Then there exists a compact element $C \in L$ such that $C \nleq B$ and $C \leq D$. For suppose $\mu \in D \sim B$, then by 5.13, (μ) is compact and clearly $(\mu) \nleq B$ but $(\mu) \leq D$. Let $C = (\mu)$.

By AL 6, there exists an $M \in L$ such that $B \leq M \leq A$, $C \nleq M$ and M is maximal. Suppose M is not meet-irreducible. Then there exists a subset $K \subseteq L$ such that $M = \bigwedge K$ and $M \notin K$. Let $N \in K$, then $M < N$. Since M is maximal $C \leq N$ for all $N \in K$. Therefore $C \leq \bigwedge K = M$. This is a contradiction, so that M is meet-irreducible.

$B \leq M$ and D is the intersection of all meet-irreducible ideals which contain B, therefore $D \leq M$. Since $C \leq D$ it follows that $C \leq M$ and this again is a contradiction. Hence $B = D$, q. e. d.

THEOREM 5.15: AL 8 $\rightarrow$ AC 3.

PROOF: Let f be an arbitrary function whose domain is s and assume for all $x \in s$, $f(x)$ has at least two elements. We define an algebra $\mathscr{A} = \langle A, \{O_u : u \in A\}\rangle$ as follows:

$$A = \{\langle x, y\rangle : x \in s \text{ and } y \in f(x)\}$$

$$O_u(v, w) = \begin{cases} u \text{ if } v = \langle x, y\rangle,\ w = \langle x, z\rangle,\ y \neq z, \\ v \text{ otherwise.} \end{cases}$$

A subset of A is a proper subalgebra if it is a function. It is clear that a subalgebra of $\mathscr{A}$ is meet-irreducible if and only if it is maximal. AL 8 implies, therefore, that there exists a maximal subalgebra $B \subset A$. If we let $g = B$, then g is a function, $\mathscr{D}(g) = \mathscr{D}(f) = s$, and for all $x \in s$, $g(x) \in f(x)$, q. e. d.

THEOREM 5.16: AL′ 3 → AL 9.

PROOF: Let A be an algebraic system. Let $D \subseteq A$ be the set of all absolutely dispensible elements and let $\mathscr{M}$ be the set of all maximal proper subalgebras. We wish to prove $D = \bigcap \mathscr{M}$.

Suppose $x \notin \bigcap \mathscr{M}$. Then there is a $M \in \mathscr{M}$ such that $x \notin M$. Therefore the smallest subalgebra containing $M \cup \{x\}$ is A, so that $x \notin D$. This implies $D \subseteq \bigcap \mathscr{M}$.

Now, suppose $x \notin D$. For any set $S \subseteq A$ let $C(S)$ be the smallest subalgebra containing S, then C is a finitary closure operator. If $x \notin D$, then there exists a set $T \subset A$ such that $C(T) \subset A$ and $C(T \cup \{x\}) = A$. We define a property of finite character P as follows: for any $S \subseteq A$, $P[S] \leftrightarrow x \notin S$. Clearly, $P[C(T)]$. Therefore AL′ 3 implies there is a maximal subalgebra M which contains $C(T)$ and $x \notin M$. Hence $x \notin \bigcap \mathscr{M}$. This implies $\bigcap \mathscr{M} \subseteq D$, q. e. d.

THEOREM 5.17: AL 10 → M 7.

PROOF: Let A be an arbitrary set and P a property of finite character. Assume A does not have the property P. For each $x \in A$ define n-ary operators on $K = \{\langle y_1, y_2, \ldots, y_n\rangle : y_1, y_2, \ldots, y_n \in A\}$ as follows:

$$\text{If } \langle y_1, y_2, \ldots, y_n\rangle \in K,\ O^n_x(y_1, y_2, \ldots, y_n) = \begin{cases} x \text{ if not} \\ \quad P[\{y_1, y_2, \ldots, y_n\}], \\ y_1 \text{ otherwise.} \end{cases}$$

$S \subset A$ is a subalgebra if and only if $P[S]$. There exists a finite set which does not have the property P; otherwise A would have the property P. Therefore, there is a smallest finite set B which does not have the property P. B is a basis for A since no proper subalgebra contains B, and every proper subset of B is a subalgebra.

Suppose $b \in B$. AL 10 implies that there exists a maximal subalgebra M such that $b \notin M$.

If $M \subset N \subseteq A$ and $P[N]$, we have $P[N \sim \{b\}]$. Since $M \subseteq N \sim \{b\}$, $N \sim \{b\} = M$ from the maximality of M. Hence $N = M \cup \{b\}$. Thus either M or $M \cup \{b\}$ is a maximal subset of A with the property P.

THEOREM 5.18: AL′ 3 $\rightarrow$ AL 11.

PROOF: Suppose A has a finite basis $B = \{b_1, b_2, \ldots, b_n\}$ and $B_0 \subset A$ is a proper subalgebra. For any set $S \subseteq A$ we define $C(S)$ as the smallest subalgebra containing S. Then C is a finitary closure operator. Define the property P_j, $j = 1, 2, \ldots, n$ as follows: for any set $S \subseteq A$, $P_j[S] \leftrightarrow b_j \notin S$. P_j is a property of finite character for each $j = 1, 2, \ldots, n$.

AL′ 3 implies that we may construct the following finite sequence of subalgebras:

If $b_j \in B_{j-1}$ then $B_j = B_{j-1}$, $j = 1, 2, \ldots, n$.

If $b_j \notin B_{j-1}$ then B_j is a maximal subalgebra containing B_{j-1} as a subset and $b_j \notin B_j$, $j = 1, 2, \ldots, n$.

By construction each B_j is a proper subalgebra and $B_{j-1} \subseteq B_j$ for each $j = 1, 2, \ldots, n$. Hence each B_j contains B_0 as a subset. We claim B_n is a maximal proper subalgebra containing B_0. For suppose B_n is not maximal. Then there exists a proper subalgebra D such that $B_n \subset D$. Since $D \neq A$, for some $b_k \in B$, $b_k \notin D$. But then $B_k \subset D$ which contradicts the definition of B_k.

THEOREM 5.19: AL 11 $\rightarrow$ M 7.

PROOF: Similar to 5.17. Take the subalgebra of AL 11 to be Λ.

6. Cardinal Number Forms

We shall discuss some propositions about cardinal numbers which are equivalent to the axiom of choice, but first we must define a cardinal number in an appropriate manner. We cannot define the cardinal number of a set x ($\overline{\overline{x}}$) to be the smallest ordinal number equivalent to x because it requires the axiom of choice to prove that such an ordinal number exists. We cannot define $\overline{\overline{x}}$ to be the class of all sets equivalent to x because in this case $\overline{\overline{x}}$ would be a proper class. What we shall do is transform the second definition by limiting the class of sets considered so that it is a set, but still obtain all the usual properties of cardinal numbers. We shall do this by introducing the notion of *rank* [1]. (Note: The axiom of regularity is needed to define rank. But rank is defined merely four convenience. Every theorem about cardinal numbers is a theorem about sets so that it would not be necessary to define cardinal numbers at all. Rank is not used in any of the proofs. However, the notion of a cardinal number greatly simplifies all of our work so we shall define it.) [2]

DEFINITION 6.1:

(1) $\mathscr{I}$ is the class of individuals.
(2) For every ordinal number α, $\tau(\alpha) = \mathscr{I} \cup \mathscr{P}(\bigcup_{\beta<\alpha} \tau(\beta))$.

LEMMA [3] 6.2: *For every x there exists an ordinal number α such that $x \in \tau(\alpha)$.*

PROOF: Suppose the lemma is false. Let $A = \{x: (\alpha)[x \notin \tau(\alpha)]\}$. Since $A \neq \Lambda$, it follows from the axiom of regularity that there exists a $y \in A$ such that $y \cap A = \Lambda$. (It is clear from the definition of τ that y is not an individual.) Since $y \cap A = \Lambda$, for every $z \in y$ there exists an α such that $z \in \tau(\alpha)$. Let $B = \{\alpha:(\exists z)[z \in y$ and $z \in \tau(\alpha)]\}$ and let $\beta = \sup B$. Since $\alpha \leq \beta$ implies $\tau(\alpha) \subseteq \tau(\beta)$, for all $z \in y$, we have $z \in \tau(\beta) = \mathscr{I} \cup \mathscr{P}(\bigcup_{\gamma<\beta} \tau(\gamma))$. Therefore, either $z \in \mathscr{I}$ or $z \subseteq \bigcup_{\gamma<\beta} \tau(\gamma)$. From this we obtain $y \subseteq \mathscr{I} \cup \mathscr{P}(\bigcup_{\gamma<\beta} \tau(\gamma)) = \tau(\beta)$ so that it follows from the definition of τ that $y \in \tau(\gamma)$ for all $\gamma > \beta$. This contradicts the assumption that $y \in A$. Consequently, the lemma follows.

DEFINITION 6.3: For every x the *rank* of x, $\rho(x)$, is the smallest ordinal number α such that $x \in \tau(\alpha)$.

[1] Defined by Mirimanoff [1], 1917. See also Tarski [8].

[2] This definition of cardinal number is due to D. Scott [2]. Scott gives credit to A. P. Morse (unpublished lecture notes) for some of the notions involved.

[3] The axiom of regularity was used in the proof.

It follows from the definitions of τ and ρ that

(1) If x is an individual, $\rho(x) = 0$,

(2) $\tau(\alpha) = \{x: \rho(x) \leq \alpha\}$.

In what follows in this section we shall have to assume that $\tau(\alpha)$ is a set. It is clear that $\tau(\alpha)$ is a set if and only if $\mathscr{I}$ is a set. Hence, we shall assume that $\mathscr{I}$, the class of individuals, is a set.

There is one further property of rank that we shall need, namely,

LEMMA 6.4: *The rank of x, $\rho(x)$, is the smallest ordinal number α such that for all $y \in x$, $\rho(y) < \alpha$.*

PROOF: Suppose $\rho(x) = \alpha$. If $x \in \mathscr{I}$ then the lemma is true. Suppose $x \notin \mathscr{I}$ and $y \in x$. Since $\rho(x) = \alpha$, $x \in \tau(\alpha) = \mathscr{I} \cup \mathscr{P}(\bigcup_{\beta<\alpha} \tau(\beta))$. Therefore, $x \subseteq \bigcup_{\beta<\alpha} \tau(\beta)$ which implies $y \in \bigcup_{\beta<\alpha} \tau(\beta)$. Hence, $y \in \tau(\beta)$ for some $\beta < \alpha$ which implies by 6.3 that $\rho(y) < \alpha$.

Now suppose $\rho(y) < \beta$ for all $y \in x$. Then, by 6.3, $y \in \bigcup_{\gamma<\beta} \tau(\gamma)$, for all $y \in x$. Consequently, $x \in \mathscr{P}(\bigcup_{\gamma<\beta} \tau(\gamma)) \subseteq \tau(\beta)$, so that $\alpha = \rho(x) \leq \beta$, q. e. d.

Now we define the cardinal number of a set as follows:

DEFINITION 6.5: The *cardinal number* of a set x, $\overline{\overline{x}}$, is the class of all sets of smallest rank which are equivalent to x.

It follows from 6.1, 6.2, and 6.3 that for every set x, $\overline{\overline{x}}$ is defined and, since $\mathscr{I}$ is a set, $\overline{\overline{x}}$ is a set. It is clear also that $x \approx y$ if and only if $\overline{\overline{x}} = \overline{\overline{y}}$.

Next, we shall define some operations and relations between cardinal numbers.

DEFINITION 6.6: Let m and n be cardinal numbers.

(1) *Addition*: Let x and y be two disjoint sets such that $\overline{\overline{x}} = m$ and $\overline{\overline{y}} = n$, then $m + n = \overline{\overline{x \cup y}}$.

(2) *Subtraction*: If there exists a unique cardinal number p such that $m = n + p$ then $m - n = p$.

(3) *Multiplication*: Let x and y be two sets such that $\overline{\overline{x}} = m$ and $\overline{\overline{y}} = n$, then $m \cdot n = \overline{\overline{x \times y}}$.

(4) *Division*: If there exists a unique cardinal number p such that $m = n \cdot p$ then $m \div n = p$.

(5) *Power*: Let x and y be two sets such that $\overline{\overline{x}} = m$ and $\overline{\overline{y}} = n$, then $m^n = \overline{\overline{\{f\colon f \text{ is a function from } y \text{ into } x\}}}$.

(6) *Inequality*: Let x and y be two sets such that $\overline{\overline{x}} = m$ and $\overline{\overline{y}} = n$, then $m < n$ if $x \prec y$ and $m \leq n$ if $x \preccurlyeq y$.

It is clear that the operations and relation as defined in 6.6 do not depend upon the particular choice of the sets x and y. Now, suppose we were to define the sum of an infinite number of cardinal numbers as follows: Let $\{A_b\colon b \in B\}$ be a class of pairwise disjoint sets then $\sum_{b \in B} \overline{\overline{A}}_b = \overline{\overline{\bigcup_{b \in B} A_b}}$. In order to show that the sum of cardinal numbers on the left side of the equality does not depend upon the particular choice of the A_b's, the axiom of choice is required. Hence, we cannot give the natural definition of an infinite sum of cardinal numbers without using the axiom of choice.

It is easy to show without the axiom of choice that for two cardinal numbers m and n, $m \leq n$ if and only if there exists a cardinal number p such that $m + p = n$. But if $+$ is changed to $\cdot$ in the preceding statement then the proposition is equivalent to the axiom of choice (see CN 12). It is also easy to show without the axiom of choice that if m and n are cardinal numbers, and $n \neq 0$ then $m \cdot n = n$ implies $m \leq n$. However, the converse of the preceding statement is equivalent to the axiom of choice, (see CN 14).

It can also be shown that addition and multiplication of cardinal numbers are commutative and associative, multiplication is distributive over addition, and the following laws hold for powers:

$$m^{n+p} = m^n \cdot m^p,$$

$$(m^n)^p = m^{n \cdot p},$$

$$m^n = m \cdot m \cdot \ldots \cdot m \ (n \text{ times}) \text{ if } n \text{ is finite.}$$

It follows from the definition of addition, multiplication, and powers that $m + n$, $m \cdot n$, and m^n exist for all cardinal numbers m and n. However, subtraction and division are different matters. For finite cardinals $n - m$ exists whenever $m < n$ and $n \div m$

exists whenever $m < n$ and n is a multiple of m. We shall show below that for all transfinite cardinals (see 6.11) m and n, if $m < n$ then $n - m$ and $n \div m$ exists if and only if the axiom of choice holds. (CN 13 and 17. See also CN 12, 14, and 18.)

Before we discuss the axiom of choice we shall first prove some preliminary lemmas

LEMMA 6.7: *If m and n are cardinal numbers and $m, n \geq 2$ then $m + n \leq m \cdot n$.*

PROOF: Let x and y be disjoint sets such that $\overline{\overline{x}} = m$ and $y = n$. Let $a, b \in x$, $c, d \in y$, $a \neq b$, and $c \neq d$. Define a mapping φ as follows:

If $s \in x$ then $\varphi(s) = \langle s, c \rangle$.

If $t \in y$ and $t \neq c$ then $\varphi(t) = \langle a, t \rangle$, and $\varphi(c) = \langle b, d \rangle$.

Clearly, φ is a 1-1 mapping of $x \cup y$ into $x \times y$. Hence $m + n \leq m \cdot n$.

DEFINITION 6.8: An *aleph* is the cardinal number of an infinite well-ordered set.

LEMMA 6.9: *For any aleph, $\aleph^2 = \aleph$.*

PROOF: Let α be an ordinal number such that $\overline{\overline{\alpha}} = \aleph$. Define a relation R on $On \times On$ as follows: For $\beta_1, \gamma_1, \beta_2, \gamma_2 \in On$, $\langle \beta_1, \gamma_1 \rangle R \langle \beta_2, \gamma_2 \rangle$ if $(\max(\beta_1, \gamma_1) < \max(\beta_2, \gamma_2)$ or $(\max(\beta_1, \gamma_1) = \max(\beta_2, \gamma_2)$ and $\beta_1 < \beta_2)$ or $(\max(\beta_1, \gamma_1) = \max(\beta_2, \gamma_2)$ and $\beta_1 = \beta_2$ and $\gamma_1 < \gamma_2)$. $On \times On$ is well-ordered by R.

Let $N = \{\beta : \beta$ is an ordinal number and (β is finite or $\overline{\overline{\beta \times \beta}} = \overline{\overline{\beta}}$)$\}$. We shall show that every ordinal number belong to N. Suppose not. Let β be the smallest ordinal number which does not belong to N. Then

(i) β is not finite.

(ii) β is a limit ordinal. For suppose not, then there exists an ordinal number γ such that $\beta = \gamma \cup \{\gamma\}$. It follows from (i) that γ is not finite, therefore $\overline{\overline{\beta}} = \overline{\overline{\gamma}}$. But, $\gamma < \beta$ so that $\gamma \in N$ which implies $\beta \in N$. This contradicts the definition of β.

(iii) $\beta \times \beta = \bigcup_{\gamma < \beta} \gamma \times \gamma$. This follows because $\gamma \times \gamma$ is an initial

R-section of $On \times On$ for every ordinal number γ and β is a limit ordinal.

Let $\psi(\gamma)$ be the ordinal number of $\gamma \times \gamma$ ordered by R. (That is, $\psi(\gamma)$ is an ordinal number and $\psi(\gamma) \cong \gamma \times \gamma$.) If $\gamma < \beta$ then

$$\overline{\overline{\psi(\gamma)}} = \overline{\overline{\gamma \times \gamma}} = \begin{cases} \text{finite} \\ \overline{\overline{\gamma}} \end{cases} < \overline{\overline{\beta}}.$$

Therefore, $\psi(\gamma) < \beta$ which implies by (iii),

$$\bigcup_{\gamma < \beta} \psi(\gamma) = \psi(\beta) \leq \beta.$$

So that $\overline{\overline{\beta \times \beta}} \leq \overline{\overline{\beta}}$ which implies $\beta \in N$. This is a contradiction, hence, $N = On$.

Lemma 6.10: *For any two alephs m and n, $m \cdot n = m + n = \max(m, n)$.*

Proof: Suppose $m \leq n$ then

$$n \leq m + n \leq m \cdot n \text{ by 6.7}$$

$$\leq n \cdot n = n^2 = n \text{ by hypotheses and 6.9, q. e. d.}$$

Definition 6.11:

(a) $\aleph_0$ is the cardinal number of the set of natural numbers.

(b) A cardinal number m is *transfinite* if $\aleph_0 \leq m$.

(c) $\aleph_\alpha$ is the αth aleph. That is, if A is the class of all alephs ordered by $<$ then $A \cong On$, where On is ordered by $<$. Then $\aleph_\alpha$ is the element of A which corresponds to $\alpha \in On$.

In all of the following propositions it is assumed that m, n, p, and q are transfinite cardinal numbers unless it is stated otherwise.

CN 1: $m \cdot n = m + n$.

CN 2: *There is a cardinal number n such that $m = n^2$.*

CN 3: $m = m^2$.

CN 4: *There is an ordinal number α such that for all transfinite cardinal numbers m there is no well-ordered by $<$ set M of cardinal numbers between m and m^2 such that $M \cong \alpha$.*

CN5 : *If* $m^2 = n^2$ *then* $m = n$.

CN 6: *If* $m < n$ *and* $p < q$ *then* $m + p < n + q$.

CN 7: *If* $m < n$ *and* $p < q$ *then* $m \cdot p < n \cdot q$.

CN 8: *If* $m + p < n + p$ *then* $m < n$.

CN 9: *If* $m \cdot p < n \cdot p$ *then* $m < n$.

CN 10: *There is a cardinal number* n *such that* $m < n$ *and for every cardinal number* p *if* $m < p$ *then* $n \leq p$. (n is said to be an *immediate successor* of m.)

CN 11: *If* $n < p$ *and there is no cardinal number between* n *and* p (p is said to *cover* n) *then either* $m \cdot n = m \cdot p$ *or* $m \cdot p$ *covers* $m \cdot n$. (If p is an immediate successor of n then p covers n. This can easily be shown without using the axiom of choice. But the converse of this is equivalent to the axiom of choice. See CN 10 and 6.19.)

CN 12: *If* $m < n$ *then there is a* p *such that* $n = m \cdot p$.

CN 13: *If* $m < n$ *then* $n \div m$ *exists.* (That is, the p of CN 12 exists and is unique.)

CN 14: *If* $m < n$ *then* $n \div m = n$.

CN 15: *If* $m + p = m + q$ *then either* $p = q$ *or* $p \leq m$ *and* $q \leq m$.

CN 16: *If* $m + m < m + n$ *then* $m < n$. (The converse is independent of the axiom of choice.)

CN 17: *If* $m < n$ *then* $n - m$ *exists.* (That is, there exists one and only one p such that $n = m + p$.)

CN 18: *If* $m < n$ *then* $n - m = n$.

(The proposition: If $m < n$ then there is a p such that $n = m + p$ (similar to CN 12) is independent of the axiom of choice.)

CN 19: *If* $p < n$ *and* $q < n$ *then* $p + q \neq n$.

CN 20: *If* $p < n$ *and* $q < n$ *then* $p \cdot q \neq n$.

CN 21: *Either* $m + n = m$ *or* $m + n = n$.

CN 22: *Either* $m \cdot n = m$ *or* $m \cdot n = n$.

Let m be a cardinal number greater than 1,

CN 23(m): *If* $p^m < q^m$ *then* $p < q$.

CN 24: *There is a cardinal number* $n > 1$ *such that for all cardinal numbers* p *and* q *there is a cardinal number* m, $1 < m \leq n$ *such hat* $p^m < q^m$ *implies* $p < q$.

CN 25: *If* $m^p < m^q$ *and* $m \neq 0$ *then* $p < q$.

CN 26: *If the greatest lower bound* $(m \cap n)$ *and the least upper bound* $(m \cup n)$ *of* m *and* n *exist then* $m \cdot n = (m \cap n) \cdot (m \cup n)$. (Tarski [7] has shown that the proposition: If $m \cap n$ and $m \cup n$ exist then $m + n = (m \cap n) + (m \cup n)$, is independent of the axiom of choice.)

In 1924, Tarski [1] proved that CN 1, CN 3, and CN 5–CN 9 are equivalent to the well-ordering theorem. CN 10 was given in 1954 by Tarski [9] and CN 11 is due to Sudan [1] (1938). CN 15–CN 19, CN 23(m), and CN 25 were stated without proof by Tarski in Lindenbaum and Tarski [1] in 1926. The proof of the equivalence of CN 17 and the well-ordering theorem was later given in 1947 by Sierpinski [4]. In 1946, CN 20 was shown to be equivalent to the well-ordering theorem by Sierpinski [2]. It was shown by M. S. Lesniewski that CN 21 implies the trichotomy and an analogous proof holds for CN 22. CN 24 is a generalization of CN 23. CN 2 and CN 12–CN 14, while not explicitly stated in the literature, follow from the work of Tarski and Sierpinski. As far as we know CN 26 is given here for the first time[1].

We shall sketch the proofs that some form of the axiom of choice implies each of the above 26 propositions.

WE 1 implies every set can be well-ordered, hence every cardinal number is an aleph. Therefore, CN 3 and CN 5 follow from 6.9, and CN 3 implies CN 2 and CN 4. CN 1 and CN 6–CN 9 follow from 6.10. Lemma 6.10 also implies CN 11–CN 22, and CN 26.

[1] A. H. Kruse has obtained several additional cardinal number forms equivalent to the axiom of choice in a paper entitled "Some observations on the axiom of choice" to appear in *Zeitschrift für Mathematische Logik und Grundlagen der Mathematik*. One form which is of particular interest because it immediately implies that both the trichotomy and the generalized continuum hypotheses imply the axiom of choise is the following: For all cardinal numbers p, m, and n, if $p < m$, $n < 2^p$ then either $m \leq n$ or $n \leq m$. See also A. H. Kruse, "Some theorems on the axiom of choice". *Notices A.M.S.* 9 (1962) 134–5 (Abstract).

CN 10 follows since the alephs are well-ordered by $<$. The trichotomy implies CN 23(m)–CN 25.

Now, in order to prove the converses, we first state and prove the following lemmas [1]:

LEMMA 6.12: *Let m, n and p be transfinite cardinal numbers, and let $\aleph$ be an aleph. If $m \cdot \aleph \leq n + p$ then either $m \leq p$ or $\aleph \leq n$.*

PROOF: Let x, y and z be sets, $y \cap z = \Lambda$, and let u be a well-ordered set such that $\overline{\overline{x}} = m$, $\overline{\overline{y}} = n$, $\overline{\overline{z}} = p$, and $\overline{\overline{u}} = \aleph$. Since $m \cdot \aleph \leq n + p$, there exists a 1-1 function φ which maps $x \times u$ into $y \cup z$.

Case 1. There is an $s \in x$ such that for all $t \in u$, $\varphi(\langle s, t\rangle) \in y$. Define $\psi(t) = \varphi(\langle s, t\rangle)$ for all $t \in u$; ψ is a 1-1 function mapping u into y, which implies $\aleph \leq n$.

Case 2. For all $s \in x$, there is a $t \in u$ such that $\varphi(\langle s, t\rangle) \in z$. Let t_s be the smallest element of u such that $\varphi(\langle s, t_s\rangle) \in z$. Define $\psi(s) = \varphi(\langle s, t_s\rangle)$ for each $s \in x$. Then ψ is a 1-1 function mapping x into z. Hence $m \leq p$, q. e. d.

LEMMA 6.13: *Let m and n be transfinite cardinal numbers and let $\aleph$ be an aleph. Then if $\aleph \leq m \cdot n$ then either $\aleph \leq m$ or $\aleph \leq n$.*

PROOF: Let x, y be sets such that $\overline{\overline{x}} = m$, $\overline{\overline{y}} = n$ and let α be a well-ordered set such that $\overline{\overline{\alpha}} = \aleph$. Then $\alpha \preccurlyeq x \times y$ so that there exist well-ordered sets $x_1, y_1, x_1 \subseteq x, y_1 \subseteq y$ such that $\alpha \preccurlyeq x_1 \times y_1$. Let $p = \overline{\overline{x_1}}$, $q = \overline{\overline{y_1}}$, then $\aleph \leq p \cdot q = \max(p, q)$ (by 6.10). If $\aleph \leq p$ then $\aleph \leq m$ and if $\aleph \leq q$ then $\aleph \leq n$, q. e. d.

DEFINITION 6.14: For any transfinite cardinal number m let $m^* = \overline{\overline{\Gamma(x)}}$ where $\overline{\overline{x}} = m$. ($\Gamma(x) = \{\alpha\colon \alpha \preccurlyeq x\}$, see 3.1, 3.2, and 3.3.)

By 3.3 we have m^* is an aleph. If $m^* \leq m$ then by 3.1 $\Gamma(x) \in \Gamma(x)$ which is impossible. Therefore, $m^* \nleq m$. Moreover, by 3.1, m^* is the smallest aleph which is not smaller than or equal to m. If m in an aleph then m^* is its immediate successor. (See CN 10 for the definition of "immediate successor".) The concept of m^* will be useful in what follows.

Remark: In the theorems that follow in which we prove that

[1] Lemma 6.12 is stated by Tarski in Lindenbaum and Tarski [1] without proof and lemma 6.13 is well known.

an arbitrary set x can be well-ordered, we assume that $m = \overline{\overline{x}}$ is transfinite. This is no loss of generality because if m is not transfinite then replace m by $m + \aleph_0$, which is transfinite. If we prove that $m + \aleph_0 \leq n$ where n is an aleph then $m \leq n$. Therefore, if α is an ordinal number such that $\overline{\overline{\alpha}} = n$ then $x \preccurlyeq \alpha$ which implies WE 3. Hence it will be sufficient to prove that for every transfinite cardinal m there exists an aleph n such that $m \leq n$.

THEOREM 6.15: CN 1 $\rightarrow$ WE 3.

PROOF: Let m be a transfinite cardinal number. By CN 1, $m \cdot m^* = m + m^*$. Since $m^* \nleq m$, we have by 6.12 that $m \leq m^*$. Since m^* is an aleph WE 3 follows.

THEOREM 6.16: CN 2 $\rightarrow$ WE 3.

PROOF: Let m be a transfinite cardinal number. CN 2 implies that there exists a cardinal number n such that $m + m^* = n^2$. Now, $m^* \leq m + m^*$, so that $m^* \leq n^2$. Since m^* is an aleph it follows from 6.12 that $m^* \leq n$. Therefore, there is a cardinal number p such that $n = m^* + p$. Now, we have,

$$\begin{aligned} m + m^* &= n^2 \\ &= (m^* + p)^2 \\ &= (m^*)^2 + 2 \cdot m^* \cdot p + p^2 \\ &\geq m^* \cdot p. \end{aligned}$$

It follows from 6.12 that either $m^* \leq m$ or $p \leq m^*$. The first alternative is impossible so that $p \leq m^*$ which implies $n = m^*$. Hence, by 6.9, $n^2 = n$, so that $m + m^* = m^*$ which implies $m \leq m^*$, q. e. d.

THEOREM 6.17: CN 3 $\rightarrow$ CN 1.

PROOF:
$$\begin{aligned} m + n &= (m + n)^2 \text{ by CN 3} \\ &= m^2 + 2 \cdot m \cdot n + n^2 \\ &\geq m \cdot n \\ &\geq m + n \text{ by 6.7.} \end{aligned}$$

Hence $m + n = m \cdot n$.

Before proving CN 4 $\rightarrow$ WE 3 we prove two more lemmas.

LEMMA 6.18: *For every transfinite cardinal m, $(m^2)^* = m^*$.*

PROOF: Let $\aleph$ be an aleph such that $\aleph \leq m \cdot m$. Then it follows from 6.13 that $\aleph \leq m$. Therefore from 6.14 it follows that $\aleph < m^*$. This implies that every aleph which is less than or equal to m^2 is also less than m^*, so that $(m^2)^* \leq m^*$. Hence $(m^2)^* = m^*$.

LEMMA 6.19: *For every transfinite cardinal m, $m < m + m^*$ and there are no cardinal numbers between m and $m + m^*$. ($m + m^*$* covers m) [1].

PROOF: For every transfinite cardinal m, $m \leq m + m^*$. If $m = m + m^*$ then $m^* \leq m$, which is impossible. Therefore $m < m + m^*$.

Suppose that p is a cardinal number such that $m \leq p \leq m + m^*$. Let $p = m + n$. We can write

(i) $m = q + r,\ n = s + t,$

where

(ii) $q + s \leq m,\ r + t \leq m^*.$

(iii) $r \leq m,\ r < m^*$ by (i), (ii).

(iv) $m + r = m$ by (iii), because (iii) implies that there exists an aleph u such that $r \leq u \leq m$. Therefore, there exists a v such that $m = u + v$. Then $m + r = u + v + r = u + v = m$.

(v) $m + s = (q + r) + s \leq m + r = m$ by (i), (ii), and (iv).

(vi) $m + n = m + t$ by (i), (v).

(vii) $t \leq m^*$ by (ii).

Therefore p can be written in the form $m + t$ where $t \leq m^*$.

Since m^* is the smallest aleph which is not less than or equal to m, and since t is an aleph, if $t < m^*$, then $t \leq m$ and $p = m + t = m$. If $t = m^*$ then $p = m + t = m + m^*$. Therefore, $m + m^*$ covers m, q. e. d.

THEOREM 6.20: CN 4 $\rightarrow$ WE 3.

PROOF: Suppose WE 3 does not hold and suppose p is a cardinal number which is not an aleph. Let $p^* = \aleph_\rho$, and let $m = p + \aleph_{\rho+\alpha}$, where α is an arbitrary ordinal number.

[1] This lemma is due to Tarski [9].

We shall show that there is a well-ordered by $<$ set M of cardinal numbers between m and m^2 such that $M \cong \alpha$. For $\beta < \alpha$, let $q_\beta = p \cdot \aleph_{\rho+\beta} + \aleph_{\rho+\alpha}$. If $\beta < \gamma < \alpha$, then $q_\beta < q_\gamma$; for suppose $q_\beta = q_\gamma$, then

$$p \cdot \aleph_{\rho+\beta} + \aleph_{\rho+\alpha} = p \cdot \aleph_{\rho+\gamma} + \aleph_{\rho+\alpha}$$

and

$$p \cdot \aleph_{\rho+\gamma} \leq p \cdot \aleph_{\rho+\beta} + \aleph_{\rho+\alpha}.$$

So by 6.12 $p \leq \aleph_{\rho+\alpha}$ or $\aleph_{\rho+\gamma} \leq p \cdot \aleph_{\rho+\beta}$.

But $p \nleq \aleph_{\rho+\alpha}$ since p is not an aleph. By 6.13, the second inequality implies that either $\aleph_{\rho+\gamma} \leq p$ or $\aleph_{\rho+\gamma} \leq \aleph_{\rho+\beta}$. The first inequality is impossible since $p^* = \aleph_\rho$ and the second is impossible since $\beta < \gamma$. Let $M = \{q_\beta \colon \beta < \alpha\}$, then $M \cong \alpha$.

Since $m = p + \aleph_{\rho+\alpha}$,

$$\begin{aligned} m^2 &= p^2 + 2 \cdot \aleph_{\rho+\alpha} \cdot p + \aleph_{\rho+\alpha}^2 \\ &= p^2 + p \cdot \aleph_{\rho+\alpha}, \end{aligned}$$

by 6.9 and other elementary properties of cardinal numbers. Therefore $m \leq q_\beta \leq m^2$ for all $\beta < \alpha$. If $m = q_\beta$ then $p + \aleph_{\rho+\alpha} = p \cdot \aleph_{\rho+\beta} + \aleph_{\rho+\beta}$. This implies that $p \cdot \aleph_{\rho+\beta} \leq p + \aleph_{\rho+\alpha}$. But by 6.12 we have either $p \leq \aleph_{\rho+\alpha}$ or $\aleph_{\rho+\beta} \leq p$ which is impossible since p is not an aleph and $p^* = \aleph_\rho$. If $q_\beta = m^2$ then $p \cdot \aleph_{\rho+\beta} + \aleph_{\rho+\alpha} = p^2 + p \cdot \aleph_{\rho+\alpha}$ which implies $p \cdot \aleph_{\rho+\alpha} \leq p \cdot \aleph_{\rho+\beta} + \aleph_{\rho+\alpha}$. Again by 6.12 either $p \leq \aleph_{\rho+\alpha}$ or $\aleph_{\rho+\alpha} \leq p \cdot \aleph_{\rho+\beta}$. The first alternative is impossible since p is not an aleph and the second alternative implies, by 6.13, that $\aleph_{\rho+\alpha} \leq p$ or $\aleph_{\rho+\alpha} \leq \aleph_{\rho+\beta}$. But both alternatives are impossible since $p^* = \aleph_\rho$ and $\beta < \alpha$. Therefore, we have $m < q_\beta < m^2$ for every $\beta < \alpha$ and since $M = \{q_\beta \colon \beta < \alpha\} \cong \alpha$, this contradicts CN 4, q. e. d.

THEOREM 6.21: CN 5 $\rightarrow$ WE 3.

PROOF: Let m be a transfinite cardinal number. There exists a cardinal number n such that $m \leq n$ and $n = n^2$. For let $n = m^{\aleph_0}$. Then $m \leq n$ and $n^2 = (m^{\aleph_0})^2 = m^{2 \cdot \aleph_0} = m^{\aleph_0} = n$.

$$(n + n^*)^2 = n^2 + 2 \cdot n \cdot n^* + (n^*)^2 \geq n \cdot n^*$$

also

$$\begin{aligned}(n + n^*)^2 &= n^2 + 2\cdot n\cdot n^* + (n^*)^2 \\ &= n + 2\cdot n\cdot n^* + n^* && \text{by hypotheses and 6.9} \\ &\leq n\cdot n^* + 2\cdot n\cdot n^* && \text{by 6.7} \\ &= n\cdot 3\cdot n^* \\ &= n\cdot n^* && \text{since } n^* \text{ is an aleph.}\end{aligned}$$

Therefore, it follows that

$$\begin{aligned}(n + n^*)^2 &= n\cdot n^* \\ &= n^2\cdot(n^*)^2 && \text{by hypotheses and 6.9} \\ &= (n\cdot n^*)^2.\end{aligned}$$

Hence by CN 5, $n\cdot n^* = n + n^*$. Since $n^* \nleq n$, we have by 6.12 $n \leq n^*$. But $m \leq n$, therefore, $m \leq n^*$ which implies WE 3, q. e. d.

THEOREM 6.22: CN 6 $\rightarrow$ WE 3.

PROOF: Let m be a transfinite cardinal number. There exists a cardinal number n such that $m \leq n$ and $2\cdot n = n$. For example let $n = \aleph_0\cdot m$. Then $m \leq n$ and $2\cdot n = 2\cdot\aleph_0\cdot m = \aleph_0\cdot m = n$. We have $n \leq n + n^*$ and $n^* \leq n + n^*$. Suppose $n < n + n^*$ and $n^* < n + n^*$. Then by CN 6,

$$n + n^* < (n + n^*) + (n + n^*) = n + n^*,$$

which is a contradiction. Therefore $n = n + n^*$ or $n^* = n + n^*$. The first alternative implies $n^* \leq n$ which is impossible. Therefore the second alternative holds and $n \leq n^*$. Hence $m \leq n^*$ and again we have WE 3, q. e. d.

THEOREM 6.23: CN 7 $\rightarrow$ WE 3.

PROOF: The proof is similar to the proof of 6.22 with $n = m^{\aleph_0} = n^2$ and replace $+$ by $\cdot$.

THEOREM 6.24: CN 8 $\rightarrow$ WE 3.

PROOF: Let m be a transfinite cardinal number. Suppose $m \nleq m^*$, then $m^* < m + m^*$. Since m^* is an aleph, $m^* + m^* = m^*$, so that $m^* + m^* < m + m^*$. Therefore, by CN 8, $m^* < m$, which is a contradiction. So that $m \leq m^*$ and WE 3 follows.

THEOREM 6.25: CN 9 $\rightarrow$ WE 3.

PROOF: The proof is analogous to the proof of 6.24 with $+$ replaced by $\cdot$.

THEOREM 6.26: CN 10 $\rightarrow$ CN 3.

PROOF: Let m be a transfinite cardinal number. CN 10 implies there is a cardinal n such that $m < n$ and for every cardinal p, if $m < p$ then $n \leq p$. But $m < m + m^*$ (by 6.19). Therefore, $m < n \leq m + m^*$. Hence, 6.19 implies $n = m + m^*$.

Suppose CN 3 does not hold. Then for some m, $m < m^2$. Since $m + m^*$ is an immediate successor of m,

(i) $m + m^* \leq m^2$,

(ii) $(m^2)^* = m^*$ by 6.18.

Therefore, (ii) implies $m^* \nleq m^2$. But (i) implies $m^* \leq m^2$. This is obviously a contradiction. Hence $m = m^2$.

THEOREM 6.27: CN 11 $\rightarrow$ WE 3.

PROOF: Let m be a transfinite cardinal number and let $m^* = \aleph_\alpha$. $\aleph_{\alpha+1}$ covers $\aleph_\alpha$, therefore, by CN 11 either $m \cdot \aleph_{\alpha+1} = m \cdot \aleph_\alpha$ or $m \cdot \aleph_{\alpha+1}$ covers $m \cdot \aleph_\alpha$. But,

$$m \cdot \aleph_\alpha \leq m \cdot \aleph_\alpha + \aleph_{\alpha+1} \leq m \cdot \aleph_{\alpha+1}.$$

Therefore, either

$$m \cdot \aleph_\alpha = m \cdot \aleph_\alpha + \aleph_{\alpha+1}$$

or

$$m \cdot \aleph_\alpha + \aleph_{\alpha+1} = m \cdot \aleph_{\alpha+1}.$$

The first alternative implies that $\aleph_{\alpha+1} \leq m \cdot \aleph_\alpha$. But $\aleph_{\alpha+1} \nleq m$ and $\aleph_{\alpha+1} \nleq \aleph_\alpha$ which contradicts 6.13. Therefore, the first alternative is impossible.

The second alternative implies either $\aleph_{\alpha+1} \leq m \cdot \aleph_\alpha$ or $m \leq \aleph_{\alpha+1}$ (6.12). It has just been shown that $\aleph_{\alpha+1} \nleq m \cdot \aleph_\alpha$, therefore, $m \leq \aleph_{\alpha+1}$ and this implies WE 3.

THEOREM 6.28: CN 12 $\rightarrow$ WE 3.

PROOF: Let m be a transfinite cardinal number. Since $m^* \leq m + m^*$, it follows from CN 12 that there is a p such that $m + m^* = m^* \cdot p$. Lemma 6.12 implies that either $m^* \leq m$ or

$p \leq m^*$. The first alternative is impossible so that $p \leq m^*$. It follows from 6.10 that $m^* \cdot p = m^*$. Therefore, $m + m^* = m^*$, which implies $m \leq m^*$, q. e. d.

Clearly CN 14 $\rightarrow$ CN 13 $\rightarrow$ CN 12.

THEOREM 6.29: CN 15 $\rightarrow$ WE 3.

PROOF: Let m be a transfinite cardinal number, then $m^* + m = m^* + (m + m^*)$. CN 15 implies either $m = m + m^*$ or $m \leq m^*$ and $m + m^* \leq m^*$. The first alternative is impossible, therefore $m \leq m^*$, q. e. d.

THEOREM 6.30: CN 16 $\rightarrow$ WE 3.

PROOF: Let m be a transfinite cardinal number, then $m^* + m^* \leq m^* + m$ since $m^* = m^* + m^*$ (6.10). If $m^* + m^* < m^* + m$ then, by CN 16, $m^* < m$, which is impossible. Hence, $m^* = m^* + m$, which implies $m \leq m^*$.

THEOREM 6.31: CN 17 $\rightarrow$ WE 3.

PROOF: Let m be a transfinite cardinal number, then $m^* \leq m + m^*$. CN 17 states that if $m^* < m + m^*$ then there is one and only one cardinal number p such that $m + m^* = m^* + p$. However, $m + m^* = m^* + m$ and $m + m^* = m^* + (m + m^*)$, (6.10). Hence, CN 17 implies $m = m + m^*$, which implies $m^* \leq m$, which is impossible. Therefore, $m^* = m + m^*$ which implies $m \leq m^*$.

It is clear that CN 18 $\rightarrow$ CN 17.

THEOREM 6.32: CN 19 $\rightarrow$ CN 18.

PROOF: Suppose $m < n$, then there is a $p \leq n$ such that $p + m = n$. CN 19 implies $p = n$. Therefore, by the definition of subtraction, $n - m = n$.

THEOREM 6.33: CN 20 $\rightarrow$ WE 3.

PROOF: Let m be a transfinite cardinal number, clearly, $m \leq m \cdot m^*$ and $m^* \leq m \cdot m^*$. CN 20 implies that either $m = m \cdot m^*$ or $m^* = m \cdot m^*$. The first alternative implies that $m^* \leq m$ which is impossible. The second alternative implies $m \leq m^*$ which implies WE 3.

THEOREM 6.34: CN 21 $\rightarrow$ T.

PROOF: If $m + n = m$ then $n \leq m$ and if $m + n = n$ then $m \leq n$, q. e. d.

THEOREM 6.35: CN 22 $\rightarrow$ T.

PROOF: The proof is analogous to the proof of 6.34 with $+$ replaced by $\cdot$.

Clearly, CN 23(m) $\rightarrow$ CN 24.

THEOREM 6.36: CN 24 $\rightarrow$ WE 3.

PROOF: Let s be a transfinite cardinal number. Let r be a cardinal number so that $r = n \cdot r$, (for example, let $r = 2^{n \cdot \aleph_0}$). Let $t = s^r$, $p = t \cdot t^*$ and $q = t + (t \cdot (t^*)^n)^*$. Suppose $1 < m \leq n$, then

$$\begin{aligned} q^m &= [t + (t \cdot (t^*)^n)^*]^m \\ &\geq t \cdot (t \cdot (t^*)^n)^{*(m-1)} \\ &\geq t \cdot (t^*)^n)^{*(m-1)} \\ &\geq t \cdot t^{*m} \; (m = m - 2 + 2 \text{ and use } 6.9) \\ &= (t \cdot t^*)^m \; (\text{by the definition of } t,\ t^m = t) \\ &= p^m. \end{aligned}$$

Next, we shall show that $p^m < q^m$. $(p^m)^* = ((t \cdot t^*)^m)^* = (t \cdot (t^*)^m)^* \leq (t \cdot (t^*)^n)^* \leq q \leq q^m$. Therefore, $p^m < q^m$ since $(p^m)^*$ is the smallest aleph which is not smaller than p^m.

Now by CN 24, it follows that $p < q$. Therefore $t \cdot t^* < t + (t \cdot (t^*)^n)^*$. By 6.12, either $t^* \leq t$ (which is impossible) or $t \leq (t \cdot (t^*)^n)^*$. Hence, $s \leq t = s^r \leq (t \cdot (t^*)^n)^*$ so that s is smaller than an aleph and that proves the theorem.

THEOREM 6.37: CN 25 $\rightarrow$ WE 3.

PROOF: Let p be a transfinite cardinal number. Let $m = 2^{p^{\aleph_0}}$ and $q = m^*$. Then

$$m^p = (2^{p^{\aleph_0}})^p = 2^{(p \cdot p^{\aleph_0})} = 2^{p^{\aleph_0 + 1}} = 2^{p^{\aleph_0}} = m \leq m^q.$$

Next, we shall show that $m^p < m^q$. $m^* = q \leq m^q$. Therefore, if $m^q \leq m^p$, $m^* \leq m$ which is impossible. Hence, $m^p < m^q$. By CN 25, $p < q = m^*$, q. e. d.

THEOREM 6.38: CN 26 $\rightarrow$ WE 3.

PROOF: Let p be a transfinite cardinal. Let $q = \aleph_\alpha$, $n = \aleph_{\alpha+1}$ where α is chosen so that $n \geq p^*$. Let $m = p + q$. Then $m \frown n = q$, $m \smile n = p + n$, $m \cdot n = p \cdot n + n$, and $(m \smile n) \cdot (m \frown n) = p \cdot q + n$. CN 26 implies $p \cdot n + n = p \cdot q + n$, which implies $p \cdot n \leq p \cdot q + n$. Therefore by 6.12 either $n \leq p \cdot q$ or $p \leq n$. The first alternative contradicts 6.13 since $n \nleq p$ and $n \nleq q$. Therefore, $p \leq n$ which implies WE 3.

7. Additional Forms

The following forms of the axiom of choice do not seem to fit into any of the earlier sections so we have made another section for them. The first three, P 1–P 3, are due to Tarski [3] and [4] and deal with sets and cardinal numbers.

P 1: *For any non-empty set s there exists a set t such that for all x, $x \in t$ if and only if $x \subseteq t$ and $\overline{\overline{s}} \nleq \overline{\overline{x}}$.*

P 2: *For any non-empty set s there exists a set t such that $t \approx \{x\colon x \subseteq t$ and $\overline{\overline{s}} \nleq \overline{\overline{x}}\}$.*

P 3: *If t is an infinite set, m and m cardinal nunbers such that $\overline{\overline{t}} = m$ and $n \leq m$, and $s = \{x\colon x \subseteq t$ and $n \nleq \overline{\overline{x}}\}$ then $\overline{\overline{s}} = m^n$.*

We shall prove the following:

$$\text{AC 3 and T} \to \text{P 1} \to \text{P 2} \to \text{WE 3},$$

$$\text{T and WE 1} \to \text{P 3} \to \text{P 2}.$$

THEOREM 7.1: AC 3 *and* T $\to$ P 1.

PROOF: Let s be a non-empty set and α a non-limit ordinal number with $(2^{\overline{\overline{s}}})^* \leq \aleph_\alpha$, (see 6.14). Let ω_α be the smallest ordinal number with cardinal number $\aleph_\alpha$. Define a function ϕ on *On* as follows:

$$\phi(\beta) = \{x\colon x \subseteq \bigcup \phi''\beta \text{ and } \overline{\overline{x}} < \overline{\overline{s}}\}.$$

Now define t as follows:

$$t = \bigcup_{\beta < \omega_\alpha} \phi(\beta) = \bigcup \phi''\omega_\alpha.$$

We shall show that t is the required set.

Suppose $x \in t$, then for some $\beta < \omega_\alpha$, $x \in \phi(\beta)$. Therefore, $x \subseteq \bigcup \phi''\beta \subseteq \bigcup \phi''\omega_\alpha = t$ and $\overline{\overline{x}} < \overline{\overline{s}}$ which implies $\overline{\overline{s}} \nleq \overline{\overline{t}}$.

Now suppose $x \subseteq t$ and $\overline{\overline{s}} \nleq \overline{\overline{x}}$. It follows from T that $\overline{\overline{s}} \nleq \overline{\overline{x}}$ if and only if $\overline{\overline{x}} < \overline{\overline{s}}$. For each $y \in x$ let $B_y = \{\beta\colon y \in \phi(\beta)\}$, let $\gamma(y)$ be the smallest ordinal number in B_y, and let $R = \mathscr{R}(\gamma)$. Since γ is a mapping of x onto R it follows from 3.5 that $\mathscr{P}(R) \preccurlyeq \mathscr{P}(x)$ or equivalently $2^{\overline{\overline{R}}} \leq 2^{\overline{\overline{x}}}$. But we also have $\overline{\overline{x}} < \overline{\overline{s}}$. Therefore, $2^{\overline{\overline{R}}} \leq 2^{\overline{\overline{s}}}$ which implies

(i) $\overline{\overline{R}} < 2^{\overline{\overline{s}}}$.

It follows from (i) that

(ii) $\overline{\overline{R}} < \aleph_\alpha$.

Since R is a set of ordinal numbers, either R is finite or $\overline{\overline{R}}$ is an aleph. If R is finite then clearly (ii) holds. If $\overline{\overline{R}}$ is an aleph, then (i) implies $\overline{\overline{R}} < (2^{\overline{\overline{s}}})^*$, so that (ii) follows from the definition of $\aleph_\alpha$.

Let $\theta = \sup R$. We claim $\theta < \omega_\alpha$. Clearly, if $\beta \in R$ then $\beta < \omega_\alpha$ which implies $\overline{\overline{\beta}} < \aleph_\alpha$. Since α is not a limit ordinal we have $\overline{\overline{\beta}} \leq \aleph_{\alpha-1}$. Let $A_\beta = \{\psi\colon \psi$ is a 1-1 function mapping β into $\omega_{\alpha-1}\}$. Now, $\overline{\overline{\beta}} \leq \aleph_{\alpha-1}$ if and only if $A_\beta \neq \Lambda$. Therefore, AC 3 implies that there exists a function f such that for each $\beta \in R$, $f_\beta \in A_\beta$. Now, we claim that

(iii) $\theta \preccurlyeq \bigcup_{\beta \in R} \{\beta\} \times \omega_{\alpha-1}$.

Clearly, $\theta \preccurlyeq \bigcup_{\beta \in R} (\{\beta\} \times \beta)$. If $\gamma \in \beta \in R$, define a function g so that $g(\langle\beta, \gamma\rangle) = \langle\beta, f_\beta(\gamma)\rangle$. Then g is a 1-1 function mapping $\bigcup_{\beta \in R} (\{\beta\} \times \beta)$ into $\bigcup_{\beta \in R} \{\beta\} \times \omega_{\alpha-1}$. Consequently (iii) follows. It follows from (iii) that

$$\begin{aligned} \overline{\overline{\theta}} &\leq \overline{\overline{R \times \omega_{\alpha-1}}} \\ &= \overline{\overline{R}} \cdot \aleph_{\alpha-1} \quad \text{by 6.6} \\ &= \aleph_{\alpha-1} \quad \text{by (ii) and 6.10.} \end{aligned}$$

Hence,

(iv) $\theta < \omega_\alpha$.

For every $y \in x$, there is a $\beta \leq \theta$ such that $y \in \phi(\beta)$. Therefore, $y \in \bigcup \phi''(\theta + 1)$, which implies $x \subseteq \bigcup \phi''(\theta + 1)$. Since $\bar{\bar{x}} < \bar{\bar{s}}$, it follows from the definition of ϕ that $x \in \phi(\theta + 1)$. Since ω_α is a limit ordinal it follows from (iv) that $\theta + 1 < \omega_\alpha$. Consequently, $x \in \bigcup \phi''\omega_\alpha = t$, q. e. d.

If in P 1 "$\bar{\bar{s}} \nleq \bar{\bar{x}}$" is replaced by "$\bar{\bar{x}} < \bar{\bar{s}}$", then the only place the axiom of choice is used to derive the resulting proposition, call it P, is in proving (iv) [1]. That is, in proving that the supremum of a set of ordinal numbers, each of which is less than ω_α, is also less than ω_α. It is known that the latter proposition is weaker than the axiom of choice, but it is not known whether P is derivable from the axioms of set theory, excluding the axiom of choice.

Before proving that P 2 $\rightarrow$ WE 3, we shall prove the following lemma:

LEMMA 7.2: *Let t and v be sets such that $\bar{\bar{v}} \leq \bar{\bar{t}}$, then there exists a set w which has the following properties:*

(i) $w \subseteq t$,

(ii) $w \notin v$,

(iii) *w can be well ordered.*

PROOF: Since $\bar{\bar{v}} \leq \bar{\bar{t}}$ there exists a 1-1 function ϕ which maps v into t. Define a function ψ such that for every ordinal number α,

$$\psi(\alpha) = \begin{cases} \phi(\psi''\alpha) & \text{if } \psi''\alpha \in v, \\ u \notin t & \text{if } \psi''\alpha \notin v. \end{cases}$$

First we note that if $\psi''\alpha$ and $\psi''\beta$ belong to v and $\psi''\alpha = \psi''\beta$ then $\alpha = \beta$. For suppose $\alpha < \beta$ then $\psi(\alpha) \in \psi''\beta$ but $\psi(\beta) \notin \psi''\beta$ so that $\psi''\alpha \neq \psi''\beta$. Since ϕ is 1-1, this implies that when the range of ψ is confined to t, ψ is 1-1. Therefore, there must be at least one ordinal number α such that $\psi''\alpha \notin v$, otherwise the set t would contain a subclass equivalent to the class of all ordinal numbers. Let γ be the smallest ordinal number in the class $\{\alpha: \psi''\alpha \notin v\}$.

Let $w = \psi''\gamma = \{\psi(\alpha): \alpha < \gamma\}$. Suppose $x \in w$. Then $x = \psi(\alpha)$ for some $\alpha < \gamma$. By the definition of γ, $\psi''\alpha \in v$, therefore $x = \psi(\alpha) = \phi(\psi''\alpha) \in t$. Hence $w \subseteq t$ and (i) is satisfied. Next, by the definition

[1] This was noted by Dana Scott.

of γ, $\psi''\gamma \notin v$, therefore $w = \psi''\gamma \notin v$ and (ii) is satisfied. Finally, when ψ is confined to $\{\alpha: \alpha < \gamma\}$, ψ is 1-1. Therefore $w = \psi''\gamma \approx \{\alpha: \alpha < \gamma\}$ so that w can be well-ordered, q. e. d.

Now we are ready to prove

THEOREM 7.3: P 2 → WE 3.

PROOF: Let s be an arbitrary set, then P 2 implies there exists a set t such that $t \approx \{x: x \subseteq t \text{ and } \bar{\bar{s}} \nleq \bar{\bar{x}}\}$. Let $v = \{x: x \subseteq t$ and $\bar{\bar{s}} \nleq \bar{\bar{x}}\}$. Since $\bar{\bar{v}} = \bar{\bar{t}}$, v and t satisfy the hypothesis of 7.2. Therefore, there exists a set w such that $w \subseteq t$, $w \notin v$ and w can be well-ordered. Since $w \subseteq t$ and $w \notin v$, we must have that $\bar{\bar{s}} \leq \bar{\bar{w}}$. Since w can be well-ordered, WE 3 follows.

THEOREM 7.4: T *and* WE 1 → P 3.

PROOF: Let t, m, and n satisfy the hypotheses of P 3. Let u be a set so that $\bar{\bar{u}} = n$.

$$s = \{x: x \subseteq t \text{ and } n \nless \bar{\bar{x}}\}$$
$$= \{x: x \subseteq t \text{ and } \bar{\bar{x}} \leq n\} \text{ by } T.$$

For each $x \in s$, let $f_x = \{f: f$ is a function from u onto $x\}$. Then

$$s \preccurlyeq \bigcup_{x \in s} f_x$$
$$\preccurlyeq \{f: f \text{ is a function from } u \text{ into } t\} = F.$$

Hence,

(i) $\bar{\bar{s}} \leq m^n$ by 6.6.

If n is finite the theorem is clear so suppose it is infinite. Then WE 1 implies that m and n are alephs so that by 6.10 $m = m \cdot n$. Therefore, there exists a 1-1 function φ which maps $t \times u$ onto t. For each $f \in F$ define a function ψ so that

$$\psi(f) = \{\varphi(f(v), v): v \in u\}.$$

ψ is 1-1 since φ is 1-1. Moreover, $\psi(f) \subseteq t$ and $\overline{\overline{\psi(f)}} = n$. Therefore ψ is a 1-1 function which maps F into s, so that

(ii) $m^n \leq \bar{\bar{s}}$.

Hence (i) and (ii) imply $\bar{\bar{s}} = m^n$.

THEOREM 7.5: P 3 → P 2.

PROOF: Let s be an arbitrary non-empty set and let $\overline{\overline{s}} = n$. Suppose $n \geq 2$. (If $n \ngeq 2$ take $t = \Lambda$ in P 2.) Let $m = 2^{n^{\aleph_0}}$; then m is infinite, $n \leq m$ and $m^n = (2^{n^{\aleph_0}})^n = 2^{n^{\aleph_0} \cdot n} = 2^{n^{\aleph_0 + 1}} = 2^{n^{\aleph_0}} = m$. Let t be a set such that $\overline{\overline{t}} = m$. P 3 implies that:

$$\begin{aligned} m^n &= \overline{\overline{\{x: x \subseteq t \text{ and } n \not\prec \overline{\overline{x}}\}}} \\ &\geq \overline{\overline{\{x: x \subseteq t \text{ and } \overline{\overline{s}} \not\preccurlyeq \overline{\overline{x}}\}}} \\ &\geq \overline{\overline{t}} = m. \end{aligned}$$

But, since $m^n = m$, we have

$$t \approx \{x: x \subseteq t \text{ and } \overline{\overline{s}} \leq \overline{\overline{x}}\}$$

and this proves P 2.

Since the trichotomy and the well-ordering theorem are equivalent it has been shown that each of P 1–P 3 is equivalent to the axiom of choice.

P 4 was stated in Lindenbaum and Tarski [1] in 1926 without proof and later was proved by Tarski [5] in 1948.

P 4: *If* $x \prec \bigcup_{i=0}^{\infty} A_i$ *then there is a finite integer* n *such that* $x \preccurlyeq \bigcup_{i=0}^{n} A_i$.

THEOREM 7.6: WE 1 $\rightarrow$ P 4.

PROOF: Suppose $x \prec \bigcup_{i=0}^{\infty} A_i$. First, it is clear that we may assume that none of the A_i's are empty. It is also clear that P 4 holds if x is finite, so let us assume x is infinite. WE 1 implies that the cardinal number of an infinite set is an aleph and also that every infinite cardinal number is transfinite.

Case 1. All the A_i's are finite. Then $\overline{\overline{\bigcup_{i=0}^{\infty} A_i}} \leq \aleph_0$ so that $x \prec \aleph_0$ which implies x is finite.

Case 2. At least one A_i is infinite, say A_j. Suppose A_k is finite, then there exists a finite set $B_k \subset A_j$ such that $A_k \approx B_k$. Since the cardinal number of A_j is transfinite, it is easy to show that $A_j \sim B_k \approx A_j$. Hence, we may assume,

Case 3. All the A_i's are infinite. Let $\aleph_\alpha = \overline{\overline{x}}$ and $\aleph_{\beta_i} = \overline{\overline{A_i}}$.

Suppose $\beta_i < \alpha$ for all i. Then $\gamma = \sup \beta \leq \alpha$ which implies

$$\overline{\overline{\bigcup_{i=0}^{\infty} A_i}} \leq \overline{\overline{\bigcup_{i=0}^{\infty} A_i \times \{i\}}}$$
$$\leq \aleph_\gamma \cdot \aleph_0$$
$$= \aleph_\gamma \text{ by } 6.10$$
$$\leq \aleph_\alpha.$$

But this contradicts the assumption that $x \prec \bigcup_{i=0}^{\infty} A_i$. Therefore, there exists an n such that $\alpha \leq \beta_n$ so that $x \preccurlyeq \bigcup_{i=0}^{n} A_i$.

THEOREM 7.7: P 4 $\rightarrow$ WE 3.

PROOF: Let x be an arbitrary set. We may assume without loss of generality that $\overline{\overline{x}}$ is transfinite. (See the remarks preceding 6.15.) We shall prove first that

(i) There exists a function F such that for each y, F_y is a 1-1 function mapping $\Gamma(y)$ into $\mathscr{P}(\mathscr{P}(\mathscr{P}(y)))$. (See 3.1, 3.2, and 3.3). Let $\alpha \in \Gamma(y)$ and let f be a 1-1 function which maps α into y. Then for any such f let $A_f = \{z : (\exists\beta)[\beta \leq \alpha \text{ and } z = f''\beta]$. If $z \in A_f$ then $z \in \mathscr{P}(y)$ so that $A_f \in \mathscr{P}(\mathscr{P}(y))$. Let $Q = \{A_f : (\exists\alpha)[\alpha \in \Gamma(y)$ and f is a 1-1 mapping of α into $y\}$. Then

(ii) $Q \subseteq \mathscr{P}(\mathscr{P}(y))$.

There is a natural mapping from Q onto $\Gamma(y)$, namely, for each $A_f \in Q$ associate the ordinal number which is the domain of f. It follows from 3.5 that there is a natural 1-1 function h which maps $\mathscr{P}(\Gamma(y))$ into $\mathscr{P}(Q)$. Define a function g from $\Gamma(y)$ to $\mathscr{P}(\Gamma(y))$ such that for each $z \in \Gamma(y)$, $g(z) = \{z\}$. Then for each $z \in \Gamma(y)$ define $F_y(z) = h(g(z))$. Since g and h are 1-1 it follows from (ii) that F_y is a 1-1 function from $\Gamma(y)$ to $\mathscr{P}(\mathscr{P}(\mathscr{P}(y)))$ which proves (i).

Now, let $y_0 = x$, $y_{i+1} = \mathscr{P}(y_i)$, and $u = \bigcup_{i=0}^{\infty} \Gamma(y_i)$. Then it follows from (i) that $u \preccurlyeq \bigcup_{i=0}^{\infty} (y_i \times \{i\})$. If $u \prec \bigcup_{i=0}^{\infty} (y_i \times \{i\})$ then by P 4 there exists a finite integer n such that $u \preccurlyeq \bigcup_{i=0}^{n} (y_i \times \{i\})$.

Therefore, there exists sets $u_i \subseteq y_i$ such that $u \approx \bigcup_{i=0}^{n} (u_i \times \{i\})$.

Then by 6.6 we have $\overline{\overline{u}} = \sum_{i=0}^{n} \overline{\overline{u_i \times \{i\}}}$. But since $\overline{\overline{u}}$ and $\overline{\overline{u_i \times \{i\}}}$ are alephs, it follows from 6.10 that $\overline{\overline{u}} = \max (\overline{\overline{u_i \times \{i\}}})$. So that there exists a j such that $\overline{\overline{u}} = \overline{\overline{u_j \times \{j\}}} = \overline{\overline{u_j}}$. This implies $u \preccurlyeq y_j$, which in turn implies the contradiction $\Gamma(y_j) \preccurlyeq y_j$.

Hence, we must have $u = \bigcup_{i=0}^{\infty} (y_i \times \{i\})$ which implies that $x = y_0 \preccurlyeq u$. But u is a set of ordinal numbers, therefore WE 3 follows.

The next proposition, P 5, is called the *Tychonoff Compactness Theorem*. It is used quite frequently in topology. In 1935 Tychonoff [1] derived P 5 from a maximal principle and later in 1950 Kelley [1] proved the converse. We shall not attempt to give these proofs here.

P5: *The product of compact spaces is compact in the product topology.*

Next, some metamathematical propositions, P 6–P 8, are considered. Vaught [2] in 1956 proved that each of them imply the axiom of choice.

P 6: *A formula having a model in a set of cardinality n also has a model in a set of cardinality m if $\aleph_0 \leq m \leq n$.*

P 7: *A formula having a model in a set of cardinality $\aleph_0$ also has a model in a set of any cardinality greater than $\aleph_0$.*

P 8: *If Q is a set of formulas in which the set of individual constants has cardinality m and every finite subset of Q has a model, then Q has a model in a set whose cardinality is not greater than $m + \aleph_0$.*

(It is assumed that the formulas contain only finitely many non-logical predicates).

We shall show that P 8 $\rightarrow$ P 7 $\rightarrow$ P 6 $\rightarrow$ CN 3.

THEOREM 7.8: P 8 $\rightarrow$ P 7.

PROOF: Let φ be a formula which has a model in a set x where $\overline{\overline{x}} = \aleph_0$. Let m be an arbitrary cardinal number such that $\aleph_0 \leq m$. Let C be a set such that $\overline{\overline{C}} = m$ and suppose $x \cap C = \Lambda$. For each $a \in C$ we define the formula φ_a as follows:

$$\varphi_a = (\exists b)[b = a].$$

For any $a_1, a_2, \ldots, a_n \in C$, φ, $\varphi_{a_1}, \varphi_{a_2}, \ldots, \varphi_{a_n}$ has a model in a set defined as follows: Let $x_1, x_2, \ldots, x_n \in x$ and suppose none of the x_i's are individual constants of φ. Replace the x_i's by the a_i's and thereby obtain a model in x.

Let $Q = \{\varphi\} \cup \{\varphi_a\colon a \in C\}$, then Q satisfies the hypothesis of P 8. Hence Q has a model in a set whose cardinal number is not greater than $m + \aleph_0 = m$. By construction of Q, the cardinal number has to be equal to m. Hence, φ has a model in a set of cardinal number m, q. e. d.

The proof that P 7 → P 6 requires the use of the well-known *Skolem-Löwenheim Theorem* (see Skolem [1]) which states:

Every consistent set of formulas (with no individual constants) has a model in a set of cardinality $\aleph_0$.

The proof of this theorem does not require the use of the axiom of choice in any form.

THEOREM 7.9: P 7 → P 6.

PROOF: Suppose a formula φ has a model in a set of cardinality n. Then φ is consistent. Therefore, by the Skolem-Löwenheim theorem, φ has a model in a set of cardinality $\aleph_0$, so that by P 7 it has a model in a set of any cardinality greater than $\aleph_0$, namely m where $\aleph_0 \leq m \leq n$, q. e. d.

THEOREM 7.10: P 6 → CN 3.

PROOF: Let m be an arbitrary transfinite cardinal number. Let $n = 2^{m \cdot \aleph_0}$, then $n^2 = (2^{m \cdot \aleph_0})^2 = 2^{m \cdot 2 \cdot \aleph_0} = 2^{m \cdot \aleph_0} = n$, and $\aleph_0 \leq m \leq n$. Consider the following formula:

S: $(x)(y)(\exists z)R(x, y, z)$ and $(x)(y)(z)(x')(y')(z')[R(x, y, z)$ and $R(x', y', z') \rightarrow (z = z' \leftrightarrow (x = x'$ and $y = y'))]$.

An algebra $\langle A, R\rangle$ where A is a set and R a ternary relation is a model for the formula S if and only if R is a 1-1 function mapping $A \times A$ into A (so that $(\overline{\overline{A}})^2 = \overline{\overline{A}}$). Clearly, the formula has a model of cardinality n since $n^2 = n$. Hence, P 6 implies it also has a model of cardinality m. Therefore $m^2 = m$, q. e. d.

The proof that the axiom of choice implies P 7 was given first in 1936 by Malcev [1] and later in 1949 by Henkin [1]. The proof will not be given here.

The propositions P 9 and P 10 are derived from a form given by von Neumann [1] [2] and they are unusual because the axiom of regularity is used to prove that the well-ordering theorem implies them. It is not known whether it can be done without this axiom.

P 9: *If x is a non-empty set and A is a proper class then A can be mapped onto x.*

P 10: *If x is a non-empty set and A is a proper class then $x \prec A$.*

Clearly P 10 implies P 9.

THEOREM 7.11: P 9 $\rightarrow$ WE 1.

PROOF: Let x be a non-empty set and let $A = On$, the class of all ordinal numbers. P 9 implies that there is a function F which maps On onto x. For each $s \in x$, $F^{-1}{}''\{s\}$ is a class of ordinal numbers. Let α_s be the smallest ordinal number in $F^{-1}{}''\{s\}$. Then x is equivalent to $\{\alpha_s : s \in x\}$. So that x is equivalent to a set of ordinal numbers and can therefore be well-ordered.

THEOREM [2] 7.12: AC 3 *and* WE 2 $\rightarrow$ P 10.

PROOF: Let x be a non-empty set and A a proper class. Let $B = \{\rho(s) : s \in A\}$ (See 6.3 for the definition of ρ. We shall assume that $\mathscr{I}$, the class of individuals, is a set). B is a proper class because A is a proper class and the class of all sets with the same rank is a set (6.3ff). Since B is a proper class of ordinal numbers it is equivalent to On. Let F be a 1-1 function mapping On onto B.

By WE 2, there is an ordinal number α, such that $x \approx \alpha$. Let ϕ be a 1-1 function which maps x onto α. By AC 3, there is a function G such that for every $\beta \in \alpha$, $G(\beta) \in F(\beta)$. Define a function ψ as follows: for each $y \in x$,

$$\psi(y) = G(\phi(y)).$$

Then ψ is a 1-1 function which maps x into A, q. e. d.

P 11 is obtained from M 14($\bar{D}$) by changing the word "set" to

[1] Von Neumann actually lists our form P 1S as an axiom of set theory. P 9 and P 10 are weaker forms.

[2] The axiom of regularity was used in the proof.

"class". (See 4.14 for the definition of $\bar{D}$.) In most cases when this is done, the new proposition is stronger than the axiom of choice. However, in this case, with the aid of the axiom of regularity, it can be shown that P 11 is equivalent to the well-ordering theorem.

P 11: *Every class has a maximal*[1] *sub-class which has the property* $\bar{D}$.

Clearly P 11 implies M 14($\bar{D}$).

THEOREM[2] 7.13: WE 2 $\to$ P 11.

PROOF: Let $X \neq \Lambda$ be an arbitrary class and let $x \in X$. WE 2 implies there exists an ordinal number α such that $x \approx \alpha$. Let ϕ be a 1-1 mapping of α onto x. Also, On is equivalent to $\alpha \times On$. (See Gödel [1], p. 27, 7.71.). Let Φ be a 1-1 mapping of On onto $\alpha \times On$, and suppose for each $\beta \in On$, $\Phi(\beta) = \langle a(\beta), b(\beta)\rangle$.

For each $\beta \in On$, define

$$S(\beta) = \{y : y \in X \text{ and } (\gamma)(z)(\gamma < \beta \text{ and } z \in S(\gamma)) \to (y \cap z \neq \Lambda \text{ and } \rho(y) = b(\beta) \text{ and } \phi(a(\beta)) \in y)\}.$$

(We assume that $\mathscr{I}$, the class of individuals, is a set so that it follows from 6.3 ff that $S(\beta)$ is a set for each $\beta \in On$.) In other words, $S(\beta)$ is the set of all elements of X which contain the $a(\beta)^{\text{th}}$ element of x, have rank $b(\beta)$, and also intersect all elements of $S(\gamma)$ for $\gamma < \beta$.

Let $Y = \bigcup_{\beta \in On} S(\beta)$, then, we say, Y is a maximal sub-class of X which has the property $\bar{D}$. Clearly, from the definition of $S(\beta)$, $Y \subseteq X$ and Y has the property $\bar{D}$ (that is, no two elements of Y are disjoint). Suppose Y is not maximal. Then there is a $z \in X$, $z \notin Y$ such that for all $y \in Y$, $z \cap Y \neq \Lambda$. Since $x \in Y$, $z \cap x \neq \Lambda$. Since $\alpha \approx x$, there is a $\beta < \alpha$ such that $\phi(\beta) \in z$, and since $On \approx \alpha \times On$, there is an ordinal number γ, such that $\Phi(\gamma) = \langle \beta, \rho(z)\rangle$. Therefore, $z \in S(\gamma)$ which implies $z \in Y$. This is a contradiction, q. e. d.

Since M 14($\bar{D}$) $\to$ M 14($\bar{D}_m$) and M 18 and M 19, it follows that if the word "set" is changed to "class" in each of M 14($\bar{D}_m$), M 18,

[1] Maximal with respect to inclusion, $\subseteq$.

[2] The axiom of regularity was used in the proof.

and M 19, the resulting propositions are equivalent to the set form of the axiom of choice.

P 12 is quite similar to AC 4. It is actually an axiom of choice for proper classes, but, since the axiom of regularity is used to prove its equivalence to AC 4, we put it in this section.

P12: *For every relation R whose domain is a set (its range may be a proper class) there is a function f such that* $\mathscr{D}(f) = \mathscr{D}(R)$ and $f \subseteq R$.

It is clear that P 12 $\rightarrow$ AC 4.

THEOREM [1] 7.14: AC 4 $\rightarrow$ P 12.

PROOF: Let R be an arbitrary relation whose domain is a set. Define a relation R' as follows:

$$R' = \{\langle x, y\rangle : \langle x, y\rangle \in R \text{ and } (z)[\langle x, z\rangle \in R \rightarrow \rho(y) \leq \rho(z)]\}.$$

(We assume that $\mathscr{I}$, the class of individuals is a set, so that it follows from 6.3ff that R' is a set.) AC 4 implies there is a function f with $\mathscr{D}(f) = \mathscr{D}(R')$ and $f \subseteq R'$. Therefore, $\mathscr{D}(R) = \mathscr{D}(f)$ and $f \subseteq R$, q. e. d.

P 13–P 15 are infinite distributive laws. P 15 is due to Collins [2] [1]. The others are generalizations of P 15.

P 13: *For every non-empty set A, for every function B whose domain is A and for each* $a \in A$, B_a *is a set, and for every function X whose domain is* $B \circ \epsilon^{-1}(\langle a, b\rangle \in B \circ \epsilon^{-1} \leftrightarrow (\exists x)[\langle a, x\rangle \in B$ *and* $b \in x])$

$$\bigcap_{a \in A} \bigcup_{b \in B_a} X_{a,b} = \bigcup_{f \in \times B_a} \bigcap_{a \in A} X_{a,f(a)}.$$

To obtain P 14 set $X_{a,b} = b$ in P 13. Slightly stronger forms, P 13s and P 14s, are obtained by allowing B to be a class-valued function in P 13 and P 14. It is immaterial in the proof whether $X_{a,b}$ in P 13 is a set or a proper class.

[1] The axiom of regularity was used in the proof.

[2] Collins uses Quine's axiom system for set theory in which the axiom of choice is inconsistent. (See Quine [1] and Specker [1].)

P 15: *For every non-empty set A*

$$\bigcap_{a \in A} \bigcup_{b \in A} b = \bigcup_{f \in C(A)} \bigcap_{a \in A} f(a)$$

where $C(A)$ is the set of all choice functions on A.

It is clear that P 13 → P 14 → P 15, P 13s → P 13, P 14s → P 14 and P 13s → P 14s.

THEOREM 7.15: P 15 → AC 3.

PROOF: Let f be an arbitrary function. (We may assume $f(x) \neq \Lambda$ for each $x \in \mathscr{D}(f)$.) For each $x \in \mathscr{D}(f)$ define $a_x = \{\{u, y\}: y \in f(x)\}$ where u is an arbitrary set. Define $A = \{a_x: x \in \mathscr{D}(f)\}$. If $b \in a_x$ for every $a_x \in A$, then $u \in b$, so that $u \in \bigcap_{a_x \in A} \bigcup_{b \in a_x} b$. Therefore, by P 15, $u \in \bigcup_{g \in C(A)} \bigcap_{a_x \in A} g(a)$. Hence, there is a $g \in C(A)$ such that for all $a_x \in A$, $u \in g(a_x)$. This implies, for all $a_x \in A$ there is an $x \in \mathscr{D}(f)$ such that $g(a_x) = \{u, y\}$ where $y \in f(x)$. Define a new function h such that $h = \{\langle x, y\rangle: x \in \mathscr{D}(f)$ and $g(a_x) = \{u, y\}\}$; then h is the required function.

THEOREM 7.16: AC 1 → P 13.

PROOF: Suppose $u \in \bigcup_{f \in \times B_a} \bigcap_{a \in A} X_{a,f(a)}$, then there is an $f \in \bigtimes_{a \in A} B_a$ such that $u \in X_{a,f(a)}$ for every $a \in A$. Therefore, for every $a \in A$ there is a $b \in B_a$, namely $b = f(a)$, such that $u \in X_{a,b}$. This implies $u \in \bigcap_{a \in A} \bigcup_{b \in B_a} X_{a,b}$. (We have shown that the right side of the equation in P 13 is a subset of the left side without using the axiom of choice.)

Next, suppose $u \in \bigcap_{a \in A} \bigcup_{b \in B_a} X_{a,b}$; then for every $a \in A$, there is a $b \in B_a$ such that $u \in X_{a,b}$. Let $C_a = \{b: b \in B_a$ and $u \in X_{a,b}\}$. By AC 1, there is a function f such that $f(C_a) \in C_a$ for all $a \in A$. This implies that $u \in \bigcup_{f \in \times B_a} \bigcap_{a \in A} X_{a,f(a)}$.

THEOREM 7.17: P 12 → P 13s.

PROOF: The proof is analogous to 7.16 until C_a is defined. In this case C_a may be a proper class. So we define a relation R such that $\langle a, b\rangle \in R \leftrightarrow a \in A$ and $b \in C_a$. P 12 implies there is a function f such that $\mathscr{D}(f) = A$ and $f \subseteq R$. Therefore, there is a function f

such that $f(a) \in C_a$, for all $a \in A$. This implies that $u \in \bigcup_{f \in \times B_a} \bigcap_{a \in A} X_{a,f(a)}$. (The axiom of regularity was used to prove AC 4 $\to$ P 12.)

THEOREM 7.18: P. 14s $\to$ P 12.

PROOF: Let R be an arbitrary relation whose domain is A. For every $a \in A$, define $B[a] = \{\{u, t\}: \langle a, t\rangle \in R\}$. For every $a \in A$, if $b \in B[a]$, then $u \in b$. Therefore, $u \in \bigcap_{a \in A} \bigcup_{b \in B[a]} b$. So by P 14s, $u \in \bigcup_{f \in \times B[a]} \bigcap_{a \in A} f(a)$. Which means that there is a function f such that for all $a \in A$, $f(a) \in B[a]$ and $u \in f(a)$. Which implies that $f(a) = \{u, t\}$ for some t such that $\langle a, t\rangle \in R$. Define $g = \{\langle a, t\rangle: \langle a, \{u, t\}\rangle \in f\}$. It is clear that g is a function. Moreover, $\mathscr{D}(g) = \mathscr{D}(R)$ and $g \subseteq R$, q. e. d.

The following implications have now been proved: AC 1 $\to$ P 13 $\to$ P 14 $\to$ P 15 $\to$ AC 3, and P12 $\to$ P 13s $\to$ P 14s $\to$ P 12. The axiom of regularity was not used in any of these proofs.

Dual forms for P 13–P 15, P 13s and P 14s can be obtained by interchanging union and intersection. We have $\bigcup_{a \in A} \bigcap_{b \in B_a} X_{a,b} = V \sim \bigcap_{a \in A} \bigcup_{b \in B_a} V \sim X_{a,b}$. However, there is a little difficulty in taking complements since the complement of a set is a proper class. But it can be shown in each case that the dual form is equivalent to the axiom of choice.

The next form is due to J. König [1]. (See also E. Zermelo [3].) [1]

P 16: *Let K be an arbitrary set and let A and B be functions with domain K. If $A_k \prec B_k$ for all $k \in K$ then* $\bigcup_{k \in K} A_k \prec \bigtimes_{k \in K} B_k$.

Clearly, P 16 $\to$ AC 6, for take $A_k = A$ for each $k \in K$.

Before proving the implication the other way, we first prove the following lemma.

LEMMA 7.19: WE 1 $\to$ *For any set K and function A whose domain is K,* $\bigcup_{k \in K} A_k \preccurlyeq \bigcup_{k \in K} \{k\} \times A_k$.

PROOF: WE 1 implies that K can be well-ordered. We define

[1] König [1] derives P 16 from the axiom of choice for the case that K is denumerable and Zermelo [3] does it for the general case.

a function f as follows: for each $x \in \bigcup_{k \in K} A_k$, $f(x) = \langle k, x \rangle$ where k is the first element of K such that $x \in A_k$. It is clear that f is a 1-1 function from $\bigcup_{k \in K} A_k$ into $\bigcup_{k \in K} \{k\} \times A_k$.

THEOREM 7.20: AC 1 *and* AC 3 *and* WE 1 $\rightarrow$ P 16.

PROOF: First, we may assume that $\overline{\overline{B}}_k > 1$ for all $k \in K$. (For suppose not. Let $J = \{k: k \in K \text{ and } \overline{\overline{B}}_k > 1\}$. Clearly $\bigcup_{k \in J} A_k = \bigcup_{k \in K} A_k$, for if $\overline{\overline{B}}_k = 1$ then $A_k = \Lambda$, and $\bigtimes_{k \in J} B_k \approx \bigtimes_{k \in K} B_k$. If $\overline{\overline{B}}_k = 1$ for all $k \in K$, P 16 holds trivially.) We may also assume that the A_k's are disjoint (by 7.19, $\bigcup_{k \in K} A_k \preccurlyeq \bigcup_{k \in K} \{k\} \times A_k$. Hence, if P 16 holds for the disjoint family $\{\{k\} \times A_k: k \in K\}$, it also holds for $\{A_k: k \in K\}$).

Since $A_k \prec B_k$, for each $k \in K$, there is a 1-1 function mapping A_k into B_k. AC 3 implies there exists a function ψ so that for each $k \in K$, $\psi(k)$ is a 1-1 function mapping A_k into B_k. Also, since $A_k \prec B_k$, $B_k \sim \psi(k) '' A_k \neq \Lambda$, AC 3 implies there exists a function ϕ such that for each $k \in K$, $\phi(k) \in B_k \sim \psi(k)'' A_k$.

Define a relation f as follows: if $x \in A_k$, j, $k \in K$,

$$f(x)(j) = \begin{cases} \phi(j) & \text{if } j \neq k, \\ \psi(k)(x) & \text{if } j = k. \end{cases}$$

Clearly, $\mathscr{D}(f) = \bigcup_{k \in K} A_k$. For each $x \in \bigcup_{k \in K} A_k$, there is one k such that $x \in A_k$ (since the A_k's are disjoint), therefore, $f(x)$ is uniquely determined. For each $j \in K$, $f(x)(j) \in B_j$. Hence, f is a function mapping $\bigcup_{k \in K} A_k$ into $\bigtimes_{k \in K} B_k$. Next, we wish to show that f is 1-1. By the definition of f, $x \in A_k$ if and only if $f(x)(k) = \psi(k)(x)$. Therefore, if $f(x) = f(y)$, $f(x)(j) = f(y)(j)$ for all $j \in K$. Hence, there exists a $k \in K$ such that $x, y \in A_k$. Moreover,

$$\psi(k)(x) = f(x)(k) = f(y)(k) = \psi(k)(y).$$

Since $\psi(k)$ is a 1-1 function, $x = y$. Hence, f is also 1-1. We have shown that $\bigcup_{k \in K} A_k \preccurlyeq \bigtimes_{k \in K} B_k$. It remains to be shown that the union and product are not equivalent.

Let $C \subseteq \mathsf{X}_{k \in K} B_k$ and $C \approx \bigcup_{k \in K} A_k$, then $C = \bigcup_{k \in K} C_k$ where the C_k's are disjoint and $C_k \approx A_k$, $k \in K$. Define $D_k = \{g(k): g \in C_k\}$, then $D_k \subseteq B_k$. Moreover, $D_k \preccurlyeq^* C_k$ (that is, there exists a function which maps C_k onto D_k), and since $C_k \approx A_k$, we also have $D_k \preccurlyeq^* A_k$. It follows from AC 1 that $D_k \preccurlyeq A_k$. Since $D_k \subseteq B_k$ and $A_k \prec B_k$, it follows that $D_k \subset B_k$. Therefore, $B_k \sim D_k \neq \Lambda$, so that AC 3 implies there exists a function h such that for each $k \in K$, $h(k) \in B_k \sim D_k$. Clearly, $h \in \mathsf{X}_{k \in K} B_k$, but by the definition of D_k, $h \notin C$. Hence, $C \neq \mathsf{X}_{k \in K} B_k$ and $\bigcup_{k \in K} A_k \prec \mathsf{X}_{k \in K} B_k$, q. e. d.

P 17 and P 18 are the next two forms to be considered in this section. We assume, for these two forms, that there are no individuals. (It could be shown that it is sufficient to assume that the class of individuals can be well-ordered.) [1] Moreover, in order to prove that these forms imply the well-ordering theorem, we use the full definition of rank (6.3).

P 17: *The power set of a well-ordered set can be well-ordered.*

P 18: *Every linearly ordered set can be well-ordered.*

We may assume without loss of generality that the well-orderings are reflexive.

Clearly WE 1 $\rightarrow$ P 17 and P 18.

THEOREM 7.21: P 18 $\rightarrow$ P 17.

PROOF: Suppose x is a well-ordered set. We define a relation R on $\mathscr{P}(x)$ as follows: if s, $t \in \mathscr{P}(x)$, $s\, R\, t \leftrightarrow$ the first element of $(s \cup t) \sim (s \cap t)$ belongs to t. $\mathscr{P}(x)$ is linearly ordered by R, hence P 18 implies $\mathscr{P}(x)$ can be well-ordered.

THEOREM [2] 7.22: P 17 $\rightarrow$ WE 1.

PROOF: We shall prove by induction that for every ordinal number α, there exists a function R such that for every $\beta < \alpha$, R_β is a reflexive well-ordering of $z(\beta) = \{x: \rho(x) = \beta\}$. (See 6.3 for the definition of rank, ρ.)

[1] There is another form of the axiom choice given by Kruse which makes this same assumption, but because of the complexity of the form we shall not give it here. (See Kruse [1] p. 552).

[2] The axiom of regularity was used in the proof.

Let $w(\beta) = \{x\colon \rho(x) < \beta\}$, and let

$$\lambda = \Gamma(w(\alpha)) \qquad \text{(see 3.1 for the definition of } \Gamma)$$
$$= \{\gamma\colon \gamma \preccurlyeq w(\alpha)\}.$$

Since λ is a set of ordinal numbers it can be well-ordered. Therefore, P 17 implies that there exists a relation S such that S well-orders $\mathscr{P}(\lambda)$. (We shall assume S is reflexive.)

$z(0) = \{\Lambda\}$ so that R_0 is well defined.

Suppose R_γ is defined for all $\gamma < \beta (< \alpha)$. Define

$$\Phi(\beta) = \bigcup_{\gamma<\beta} R_\gamma \cup \bigcup_{\gamma<\delta<\beta} z(\gamma) \times z(\delta).$$

$\Phi(\beta)$ is a well-ordering of $w(\beta)$. For each reflexive well-ordering P, $\mathscr{D}(P)$ is equivalent to an ordinal number and there is just one 1-1 function Ψ_P which maps $\mathscr{D}(P)$ onto that ordinal number preserving order. (See Gödel [1] p. 27). Then $\Psi_{\Phi(\beta)}$ is a 1-1 mapping of $w(\beta)$ onto an ordinal number. We have

$$\Psi_{\Phi(\beta)}(x) \prec \Psi_{\Phi(\beta)}(y) \leftrightarrow x\Phi(\beta)y,$$

and

$$\Psi_{\Phi(\beta)}{}''w(\beta) \prec \lambda.$$

It follows from 6.4 that if $x, y \in z(\beta)$ then $x, y \in \mathscr{P}(w(\beta))$ so that we define R_β as follows:

$$x\, R_\beta\, y \leftrightarrow \Psi_{\Phi(\beta)}{}''x\; S\; \Psi_{\Phi(\beta)}{}''\, y.$$

Then clearly R_β is a reflexive well-ordering of $z(\beta)$, q. e. d.

P 19 is put in primarily as a curiosity. Blair and Tomber [1] discuss two forms of the axiom of choice for finite sets. It was pointed out to them by H. Rubin that if a slight change was made in the wording of these two forms, one would obtain two forms of the general axiom of choice. (This is noted in Blair and Tomber's paper.) P 19 is a generalization of these two forms.

P 19: *Let R be a transitive, anti-symmetric relation on a non-empty set x. Consider the set $z_t = \{s\colon s \in x$ and $t \in x$ and $s \neq t$ and $t\, R\, s\}$. Suppose, for each $t \in x$, there is an R-minimal element $u \in z_t$ such that $u\, R\, s$ for each $s \in z_t$, $u \neq s$. Then there exists an R-linearly ordered subset of x with no R-upper bound.*

We shall show that P 19 is equivalent to M 1.

THEOREM 7.23: M 1 $\rightarrow$ P 19.

PROOF: Let x be an arbitrary set and R a transitive relation on x. Suppose x has an R-maximal element m. Then $z_m = \Lambda$ and, hence, the hypothesis of P 19 fails to be true. Therefore, if the hypothesis of P 19 is true then x has no R-maximal element. So that the contrapositive of M 1 implies that x has an R-linearly ordered subset with no R-upper bound, q. e. d.

THEOREM 7.24: P 19 $\rightarrow$ M 1.

PROOF: Suppose M 1 is false. Then there exists a set x and a transitive, anti-symmetric relation R defined on x (we can assume R is anti-symmetric without loss of generality) such that every R-linearly ordered subset of x has an R-upper bound and x has no R-maximal element. Let N be the set of integers. We define a relation S on $x \times N$ as follows: if $\langle s, m\rangle$, $\langle t, n\rangle \in x \times N$ then $\langle s, m\rangle \, S \, \langle t, n\rangle$ if $s \neq t$ and $s \, R \, t$ or $s = t$ and $m < n$. $x \times N$ satisfies the hypothesis of P 19, because if $\langle s, m\rangle \, S \, \langle t, n\rangle$ then $\langle s, m\rangle \, S \, \langle s, m+1\rangle$ and $\langle s, m+1\rangle \, S \, \langle t, n\rangle$ or $\langle s, m+1\rangle = \langle t, n\rangle$. Hence, $\langle s, m+1\rangle$ is an S-minimal element of $z_{\langle s, m\rangle}$ such that for all $\langle t, n\rangle \in z_{\langle s, m\rangle}$, $\langle s, m+1\rangle \neq \langle t, n\rangle$, $\langle s, m+1\rangle \, S \, \langle t, n\rangle$.

Let u be an S-linearly ordered subset of $x \times N$, and let $y = \mathscr{D}(u)$. Then y is an R-linearly ordered subset of x. Every R-linearly ordered subset of x has a strict R-upper bound. If not then an R-upper bound would be an R-maximal element of x. Let b be a strict R-upper bound of y. Then $\langle b, 0\rangle$ is an S-upper bound of u. This contradicts P 19, q. e. d.

P 20, which is the last form to be considered in this section, is due to Tarski [8].

P 20: *It is not the case that there exists a set x such that if $y = \mathscr{P}(x) \sim \{\Lambda\}$ and $F = \{f: f$ is a function mapping y into $x\}$ then there exists a function g such that $\mathscr{D}(g) = F$, $\mathscr{R}(g) \subseteq y$ and $f(g(f)) \notin g(f)$ for all $f \in F$.*

THEOREM 7.25: AC 1 $\rightarrow$ P 20.

PROOF: Suppose P 20 is false. Then there is a set x and a function g such that $\mathscr{D}(g) = F$, $\mathscr{R}(g) = s \subseteq y$, and $f(g(f)) \notin g(f)$ for all $f \in F$. This implies that s is a set for which AC 1 does not hold, q. e. d.

THEOREM 7.26: P 20 $\rightarrow$ AC 1.

PROOF: Suppose AC 1 is false. Then there is a set x such that for every $f \in F$, there is a $z \in y$ such that $f(z) \notin z$. Suppose $u \notin x$; then define

$$h_f(\alpha) = \begin{cases} f(x \sim h_f{}''\alpha) & \text{if } f(x \sim h_f{}''\alpha) \in x \sim h_f{}''\alpha, \\ u & \text{otherwise.} \end{cases}$$

Just as in the proof of 2.8 (for G) we can prove that h_f^{-1} is 1-1 on $\mathscr{R}(h_f) \cap x$ and therefore there exists an α such that $h_f(\alpha) = u$. Let α be the smallest ordinal number such that $\alpha \in \{\beta\colon h_f(\beta) = u\}$. If $h_f{}''\alpha = x$, then it follows, just as in 2.8, that x can be well-ordered. But then, it is clear that we can find a function $f \in F$ such that for all $z \in y$, $f(z) \in z$ (just take $f(z)$ to be the first element of z) so this contradicts our assumption that AC 1 is false. Therefore, $h_f{}''\alpha \subset x$, so define $g(f) = x \sim h_f{}''\alpha$. Then g has the property that $\mathscr{D}(g) = F$, $\mathscr{R}(g) \subseteq y$ and $f(g(f)) = u \notin g(f)$, so that this contradicts P 20, q. e. d.

In conclusion we shall summarize the work in section 7 by indicating where the axiom of regularity was used in proving equivalences. All the propositions given in sections 1–6 were shown to be equivalent without using the axiom of regularity. In section 7 the following propositions were shown to be equivalent without using the axiom of regularity:

A: P 1–P 8, P 13–P 16, P 19, P 20, AC 1 (all forms given in sections 1–6).

B: P 12, P 13s, P 14s.

In the following diagram when we write for example B $\rightarrow$ A, we mean one of the equivalent propositions in B implies one of the equivalent propositions in A.

P 11 $\searrow$

P 10 $\rightarrow$ P 9 $\rightarrow$ A $\rightarrow$ P 18 $\rightarrow$ P 17.

B $\nearrow$

B and P 9 $\rightarrow$ P 10.

Using the axiom of regularity and that the class of individuals is a set, it was shown that A → P 11, A → P 10, and A → B. (It could have been shown that it was sufficient to assume the axiom of regularity and either the class of individuals is a set or the class of individuals can be well-ordered.) Using the axiom of regularity and the class of individuals can be well-ordered, it was shown that P 17 → A. (In proving the former implications the only property of rank that was needed was the class of all sets with the same rank is a set while in proving the latter implication other properties of rank were also used.)

PART II

CLASS FORMS

We now turn to the class forms of the axiom of choice. There is not very much in the literature on the class forms since they are hardly ever used in practice. In 1925 J. von Neumann [1] listed P 1S as an axiom of set theory (Axiom IV 2) and stated that the well-ordering theorem can be derived from it. In 1928, von Neumann [2] proved that his Axiom IV 2 implies that the universe can be well-ordered. In 1940, Gödel [1] used AC 1S as an axiom of set theory [1] and Bernays [2], in 1941, used AC 4S as an axiom. Bernays also proved the equivalence of AC 4S and AC 5S.

For each set form of the axiom of choice in Part I which has the form – "For every x there exists a y such that $P(x, y)$", where P is a property – we can construct the class form – "There is a function F such that for every x, $P(x, F(x))$". For an example of this see WE 1 and WE 3S. Another example is a class form of T:

There is a function F such that for every two sets x and y, $F(x, y)$ is a 1-1 function of a subset of x into y such that either $\mathscr{D}(F(x, y)) = x$ or $\mathscr{R}(F(x, y)) = y$.

The proofs that all class forms of this type are equivalent are similar to the corresponding proofs for the set forms. It is clear that class forms of this type imply the corresponding set forms. We shall not give all such class forms here.

In many cases a class form can be obtained from a set form merely by changing the word "set" to "class". (For example, see AC 1 and AC 1S, AC 2 and AC 2S, AC 3 and AC 3S, AC 4, P 12, and AC 4S, etc.) In other cases other changes have to be

[1] Actually, Gödel stated the following form as an axiom: There is a function F such that for every non-empty set x, $F(x) \in x$. (Axiom E). But this is clearly equivalent to AC 1S.

made also. (See, for example, M 1S–M 4S). In some cases there are other methods for strengthening set forms. (See, for example, MR 6S and MF 3S.) There are also some class forms which have no analogue in Part I (MR 14S(U) and MR 14S($\overline{U}$). We shall not attempt to give all the class forms which are obtainable from the set forms of Part I.

1. The Well-Ordering Theorem [1]

WE 1S–4S are class forms of WE 1, WE 5S is a class form of WE 2, and WE 6(m)S–WE 8S are class forms of WE 4(m)–WE 6.

WE 1S: *There is a relation R such that every set is well-ordered by R.*

WE 2S: *There is a relation R such that every class is well-ordered by R.*

WE 3S: *There is a function F such that every set x is well-ordered by F(x).*

WE 4S: *There is a relation R such that R well-orders V and every proper initial R-section of V is a set.*

WE 5S: *If X is a proper class then X is equivalent to On.*

Let m be a natural number, $m \geq 1$.

WE 6(m)S: *For every class X there exists a function F defined on On such that* $F(\alpha) \preccurlyeq m$ *for every* $\alpha \in On$ *and* $\bigcup_{\alpha \in On} F(\alpha) = X$.

WE 7S: *There exists a natural number* $m \geq 1$, *such that* WE 6(m)S.

WE 8S: *For every class X there exists a natural number* $m \geq 1$ *and a function F defined on On such that* $F(\alpha) \preccurlyeq m$ *for every* $\alpha \in On$ *and* $\bigcup_{\alpha \in On} F(\alpha) = X$.

The following implications are immediate:

WE 5S $\to$ WE 2S $\to$ WE 1S $\to$ WE 3S, WE 5S $\to$ WE 4S,

WE 6(1)S $\leftrightarrow$ WE 5S, WE 6(m)S $\to$ WE 6(n)S if $m \leq n$, and

[1] The letter "S" will follow the number of each proposition, definition and theorem of Part II.

WE 6(m)S $\rightarrow$ WE 7S $\rightarrow$ WE 8S.

THEOREM 1.1S: WE 4S $\rightarrow$ WE 5S.

PROOF: Use Theorem 7.7 in Gödel [1], p. 27.

THEOREM [1] 1.2S: WE 3S $\rightarrow$ WE 4S.

PROOF: Suppose F satisfies WE 3S. Define R as follows: $x\,R\,y \leftrightarrow [(\rho(x) < \rho(y)$ or $(\rho(x) = \rho(y)$ and x precedes y in the well-ordering of $\{z: \rho(z) = \rho(x)\}$ by $F(\{z: \rho(z) = \rho(x)\})]$. (See 6.3 ff.)

R is clearly an ordering of the universe. Moreover, if x is a set, $\{y: y\,R\,x\} \subseteq \{y: \rho(y) \leq \rho(x)\}$. We shall assume that $\mathscr{I}$, the class of individuals, is a set so that the latter class is also a set. Hence, every proper initial R-section of V is a set.

It is also clear that R is connected. Therefore, to prove R well-orders V, it is sufficient to show that every non-empty subclass of V has an R-first element.

Let $A \subseteq V$ and $A \neq \Lambda$. The elements of A are partially ordered by their rank. Let α be the smallest ordinal number which is the rank of an element in A and let $B = \{x: x \in A$ and $\rho(x) = \alpha\}$. B is a set since $\mathscr{I}$ is a set. B is well-ordered by R by construction. Let b be the R-first element of B. Then b is also the R-first element of A, q. e. d.

THEOREM 1.3S: WE 8S $\rightarrow$ WE 5S.

PROOF: The proof is similar to 1.1, but here it is sufficient to take the universe, V, as the proper class and therefore, we automatically have $V \times V \subseteq V$. Here (i) becomes

(i′) $m \in N$ if $(\exists F)[F$ is a function, $\mathscr{D}(F) = On$, $\bigcup_{\alpha \in On} F(\alpha) = V$, and $(\alpha)[\alpha \in On \rightarrow F(\alpha) \preccurlyeq m]]$ and (ii) becomes

(ii′) If m is a natural number such that $m > 1$ then $m \in N$ implies $m - 1 \in N$. The proof goes through just as in 1.1.

It has now been shown that WE 1S–WE 8S are all equivalent.

2. The Axiom of Choice

Forms AC 1S–AC 5S and AC 7(m)S–AC 9S may be obtained from the corresponding set forms by substituting the word "class"

[1] The axiom of regularity was used in the proof.

for "set" in the appropriate places. AC 6 has no class analogue because the cartesian product of a proper class of non-empty sets is empty. (This was noted by A. Tarski.)

AC 1S: *If S is a class of non-empty sets, there is a function F such that for each $x \in S$, $F(x) \in x$.*

AC 2S: *If T is a disjoint class of non-empty sets, there is a class C which consists of one and only one element from each set in T.*

AC 3S: *For every function F there is a function G such that for every x if $x \in \mathscr{D}(F)$ and $F(x) \neq \Lambda$ then $G(x) \in F(x)$.*

AC 4S: *For every relation R there is a function F such that $\mathscr{D}(F) = \mathscr{D}(R)$ and $F \subseteq R$.* (See P 12.)

AC 5S: *For every function F there is a function G such that $\mathscr{D}(G) = \mathscr{R}(F)$ and for every $x \in \mathscr{D}(G)$, $F(G(x)) = x$.*

Let m be a natural number, $m \geq 1$.

AC 7(m)S: *If S is a class of non-empty sets, there is a function F such that for every $x \in S$, $F(x) \neq \Lambda$, $F(x) \subseteq x$, and $F(x) \preccurlyeq m$.*

AC 8S: *There exists a natural number $m \geq 1$ such that* AC 7(m)S.

AC 9S: *If S is a class of non-empty sets, then there is a natural number $m \geq 1$ and a function F such that for every $x \in S$, $F(x) \neq \Lambda$, $F(x) \subseteq x$, and $F(x) \preccurlyeq m$.*

The proofs that AC 1S–AC 5S are equivalent are the same as the corresponding proofs for the set forms with one exception. The proof that AC 3 $\rightarrow$ AC 4 (see 2.7) does not hold for the class forms because for some $x \in \mathscr{D}(R)$ (where R is the relation described in AC 4S), $\{y: \langle x, y\rangle \in R\}$ may be a proper class.

To prove AC 3S $\rightarrow$ AC 4S we introduce the following principle which is weaker than the axiom of regularity or the existence of rank.

PR: *For every relation R there exists a relation S such that $\mathscr{D}(R) = \mathscr{D}(S)$, $S \subseteq R$ and for every x, $S''\{x\}$ is a set.*

LEMMA 2.1S: *The axiom of regularity and $\mathscr{I}$ is a set imply PR.*

PROOF: Let R be an arbitrary relation. Define S as follows: $x\, S\, y \leftrightarrow x\, R\, y$ and $(z)[x\, R\, z \rightarrow \rho(y) \leq \rho(z)]$. It follows from 6.3 ff that $S''\{x\}$ is a set for all x.

It is clear that AC 4S also implies PR.

THEOREM 2.2S: *PR and* AC 3S $\to$ AC 4S.

PROOF: Let R be an arbitrary relation. PR implies that there exists a relation S such that $\mathscr{D}(R) = \mathscr{D}(S)$, $S \subseteq R$ and for every x, $S''\{x\}$ is a set. For each $x \in \mathscr{D}(S)$ define a function F as follows: $F(x) = S''\{x\}$. AC 3S implies there is a function G such that for every x, if $x \in \mathscr{D}(F)$ and $F(x) \neq \Lambda$ then $G(x) \in F(x)$. G is the required function, q. e. d.

We have now shown that AC 1S–AC 5S are all equivalent. We shall show next that they are also equivalent to the well-ordering theorem.

Clearly WE 3S $\to$ AC 1S.

THEOREM 2.3S: AC 1S $\to$ WE 3S.

PROOF: AC 1S $\to$ AC 1 $\to$ WE 1. For every non-empty set x, let

$$S_x = \{R: R \subseteq x \times x \text{ and } R \text{ well-orders } x\}.$$

WE 1 $\to S_x \neq \Lambda$. Let $S = \{S_x: x \in V\}$. By AC 1S there is a function G such that for each $S_x \in S$, $G(S_x) \in S_x$. Then, for each $x \in V$, define $F(x) = G(S_x)$. F is the required function.

Also, without using the axiom of regularity we can prove the following:

THEOREM 2.4S: WE 2S $\to$ AC 4S.

PROOF: Let R be an arbitrary relation. By WE 2S there exists a relation S, such that for every class X, X can be well-ordered by S. Define a function F as follows. If $x \in \mathscr{D}(R)$ then $F(x) = S$-first element of $\{y: x\, R\, y\}$. F is the required function.

Just as in the case of the corresponding set forms of the axiom of choice the following equivalences are clear: AC 1S $\leftrightarrow$ AC 7(1)S, AC 7(m)S $\to$ AC 7(n)S if $m \leq n$, and AC 7(m)S $\to$ AC 8S $\leftrightarrow$ AC 9S. So that now it remains to be shown that AC 9S implies a form of the well-ordering theorem.

THEOREM[1] 2.5S: AC 9S $\to$ WE 8S.

PROOF: (The proof is similar to 2.9 but here we have to use the axiom of regularity and assume $\mathscr{I}$ is a set).

[1] The axiom of regularity was used in the proof.

Let S be the class of all non-empty sets. Then by AC 9S there is a natural number $m \geq 1$, and a function F such that for every $x \in S$, $F(x) \neq \Lambda$, $F(x) \subseteq x$, and $F(x) \preccurlyeq m$. Define a function G on On as follows:

$$G(\alpha) = F(\{x: x \notin \bigcup G''\alpha \text{ and } (y)[y \notin \bigcup G''\alpha \rightarrow \rho(x) \leq \rho(y)]\}).$$

Then just as in 2.9 we can prove

(1) G^{-1} is 1-1 on $\mathscr{R}(G) \cap S$ and

(2) $\bigcup G'' \, On = V$.

Now, WE 8S follows.

Before concluding this section we shall mention another proposition which is clearly weaker than the axiom of choice, but with its help we can prove WE 1S implies WE 2S. We state it here, rather than in section 1, because it is a direct consequence of AC 4S. It is called the Principle of Dependent choices (PD) and is due to Tarski [5].

PD: *For every non-empty relation R, if $\mathscr{R}(R) \subseteq \mathscr{D}(R)$ then there is a function f, $\mathscr{D}(f) =$ set of natural numbers, such that for each natural number n, $f(n) \, R \, f(n+1)$.*

THEOREM 2.6S: AC 4S $\rightarrow$ PD.

PROOF: Let R be a relation which satisfies the hypotheses of PD. By AC 4S, there is a function G, such that $\mathscr{D}(G) = \mathscr{D}(R)$ and $G \subseteq R$. Let $x \in \mathscr{D}(R)$. Define

$$f(0) = x,$$
$$f(m+1) = G(f(m)).$$

Then f is clearly the required function.

THEOREM 2.7S: WE 1S *and* PD $\rightarrow$ WE 2S.

PROOF: Let W be an irreflexive relation which well-orders every set. Suppose W does not well-order the universe. Then there is a non-empty class A such that A has no W-first element. Define a relation R as follows:

$R = \{\langle x, y\rangle : x \in A, y \in A \text{ and } \langle y, x\rangle \in W\}$. Since A has no W-first element, $\mathscr{D}(R) = A$. Therefore, R satisfies the hypotheses of PD. Hence, there is a function f such that for each natural number n, $\langle f(n), f(n+1)\rangle \in R$. But, by the definition of R, this implies that $\langle f(n+1), f(n)\rangle \in W$ for each n. So that the set $\{f(n)$: n is a natural number$\}$ has no W-first element. This is a contradiction, q. e. d.

3. Maximal Principles

M 1S–M 4S are class forms of M 1–M 4.

M 1S: *If R is a transitive relation on a non-empty class X and if every subset of X which is linearly ordered by R has an R-upper bound then either there is an R-maximal element in X or X has a subclass which is a proper class and which is linearly ordered by R.*

M 2S: *If R is a transitive relation on a non-empty class X and if every subset of X which is well-ordered by R has an R-upper bound then either there is an R-maximal element in X or X has a subclass which is a proper class and which is well-ordered by R.*

M3S: *If every non-empty nest which is a subset of a non-empty class X has its union in X, then either X has a maximal element*[1] *or there is a subclass of X which is a nest and a proper class.*

M 4S: *If every non-empty well-ordered nest which is a subset of a non-empty class X has its union in X, then either X has a maximal element*[1] *or there is a subclass of X which is a well-ordered nest and a proper class.*

Consider the following set forms: M 5–M 13, M 14(D), ($\bar{J}$), (K), (M 14(K) = M 6), M 14(D_m), M 16, M 17, and M 23. If, in each of these forms the word "set" is changed to "class", we obtain class forms, M 5S–M 13S, M 14S(D), ($\bar{J}$), (K), (M 14S(K) = M 6S), M 14S(D_m), M 16S, M 17S, and M 23S. If the word "set" is changed to "class" in M 14($\bar{D}$) we obtain P 11 which is equivalent to a set form. The same is true for M 14($\bar{D}_m$), M 18, and M 19. If in M 14(J), the word "set" is changed to "class", we do not know if the resulting proposition is equivalent to any form of the axiom of choice. However, we were able to prove that the following propositions are equivalent to the class form of the axiom of choice when A is replaced by J or $\bar{K}$.

[1] Maximal with respect to inclusion, $\subseteq$.

M 15S(A): *The universe can be anti-symmetrically linearly ordered and every class has a maximal subclass which has the property* A.

As was pointed out in Part I (preceding Definition 4.14), if X and Y are distinct, non-empty sets then exactly one of the following properties hold: $X\ D\ Y$, $X\ K\ Y$, or $X\ J\ Y$. However, in the case X and Y are classes there is another possibility to consider, namely $X \cup Y = V$.

DEFINITION 3.1S: Let X and Y be classes;

(i) $X\ U\ Y$ if $X \cup Y = V$,

(ii) $X\ \overline{U}\ Y$ if $X \cup Y \neq V$.

The property U is dual to D. (That is, $X\ D\ Y \leftrightarrow (V \sim X)\ U\ (V \sim Y)$). If X and Y are distinct, non-empty, non-universal, and non-complementary classes then exactly one of the properties D, K, J, and U hold between them. $\overline{D}$, $\overline{K}$, $\overline{J}$, and $\overline{U}$ are the complements of D, K, J, and U respectively, therefore, we must have $\overline{D} = K$ or J or U, $\overline{K} = D$ or J or U, $\overline{J} = D$ or K or U, and $\overline{U} = D$ or K or J. For the sake of completeness we shall also consider the six properties, D or K (DK), D or J (DJ), D or U (DU), K or J (KJ), K or U (KU), and J or U (JU), which in this case are not the complements of D, K, J or U.

It would be impossible to state the form M 14S(A) where U or $\overline{U}$ is substituted for A because it would be necessary to construct a class of classes. In order to avert this difficulty we introduce a relation in an appropriate manner.

MR 14S(A): *For every relation* R *and class* X *there exists a maximal subclass* [1] Y *such that for all* $x, y \in Y$, $x \neq y$, $R''\{x\}\ A\ R''\{y\}$.

MR 14S(A) is equivalent to the class form of the axiom of choice if $A = D$, $\overline{D}$, J, $\overline{J}$, K, D_m, $\overline{D}_m$, U, $\overline{U}$, DK, DJ, DU, KJ, KU, or JU. (D_m and $\overline{D}_m$ are not binary properties so that by MR 14S(D_m) we mean the following:

For every relation R and class X there exists a maximal sub-

[1] Maximal with respect to inclusion, $\subseteq$.

class Y such that for all $y_1, y_2, \ldots, y_m \in Y$, y_i's distinct, $\bigcap_{i=1}^{m} R''\{y_i\} = \Lambda$. Similarly for $\bar{D}_m$.)

The following propositions are also equivalent to the class form of the axiom of choice.

MR 15S($\bar{K}$): *The universe can be anti-symmetrically linearly ordered and for every relation R and class X there exists a maximal subclass*[1] *Y such that for all $x, y \in Y$, $x \neq y$, $R''\{x\}\ \bar{K}\ R''\{y\}$.*

MR 16S: *There exists a natural number $m \geq 2$ such that* MR 14S(D_m).

MR 17S: *For every relation R and class X, there exists a natural number $m \geq 2$ and a maximal subclass*[1] *Y such that for all $y_1, y_2, \ldots, y_m \in Y$, y_i's distinct, $\bigcap_{i=1}^{m} R''\{y_i\} = \Lambda$.*

MR 18S: *There exists a natural number $m \geq 2$ such that* MR 14S($\overline{D!_m}$).

MR 19S: *For every relation R and class X, there exists a natural number $m \geq 2$ and a maximal subclass*[1] *Y such that for all $y_1, y_2, \ldots, y_m \in Y$, $\bigcap_{i=1}^{m} R''\{y_i\} \neq \Lambda$.*

While we were not able to prove M 14S(J) equivalent to the other class forms we are able to prove MR 14S(J) equivalent to a class form. Even though M 14S($\bar{D}$), M 18S, and M 19S are equivalent to a set form, MR 14S($\bar{D}$), MR 18S, and MR 19S are equivalent to a class form.

The following proposition is put in primarily for convenience in proving certain equivalences.

MF 3S: *Let F be a class valued function, Π a predicate such that for each class X,*

(i) $\Pi[X] \leftrightarrow (x)[x \subseteq X \rightarrow x \in \mathscr{D}(F)$ *and* $F[x] \subseteq X]$

(ii) $x \subseteq F[x]$ *for all* $x \in \mathscr{D}(F)$ *and*

(iii) *if for every non-empty nest* $y \subseteq \mathscr{D}(F)$, $\Pi[\bigcup_{y \in Y} F[y]]$, *then if*

[1] Maximal with respect to inclusion, $\subseteq$.

there is a class which has the property Π, then there is a maximal class[1] *which has the property Π.*

There is also a form MF 4S which is obtained from MF 3S merely by requiring the nest to be well-ordered. MF 3S and MF 4S are class forms of M 3 and M 4.

The final maximal principle to be considered in this section is AL′ 3S which is obtained from AL′ 3 by changing the word "set" to "class" wherever it occurs. We shall not consider the class forms corresponding to the other set forms in sections 5 and 6.

We shall now proceed to prove that all these propositions are equivalent. In many cases the proofs of equivalence for the set forms carry over for the corresponding class forms. However, in some cases the proofs do not carry over because the constructions would lead to a class of classes for the class forms. It is usually possible to avoid this difficulty by using the notion of rank and we shall indicate how this is done in the theorems that follow.

We shall first prove the following:

(I) M 5S $\leftrightarrow$ MR 14S(K) $\rightarrow$ M 6S $\leftrightarrow$ M 1S $\leftrightarrow$ M 2S $\leftrightarrow$ M 3S $\leftrightarrow$ M 4S $\leftrightarrow$ M 23S.

(Note: M 6S is the same as M 14S(K).)

Clearly M 2S $\rightarrow$ M 1S $\rightarrow$ M 3S and M 2S $\rightarrow$ M 4S $\rightarrow$ M 3S.

THEOREM 3.2S: M 3S $\rightarrow$ M 2S.

PROOF: Let X be an arbitrary class and R a transitive relation on X. Define Y as follows:

$Y = \{s\colon s \subseteq \mathscr{P}(X)$ and $(x)[x \in s \rightarrow R$ is a well-ordering on $x]$ and $(x)(y)[x, y \in s \rightarrow x$ is an initial R-section of y or y is an initial R-section of $x]\}$.

Let $n \subseteq Y$ be a nest, then $n \in Y$. Hence, by M 3S, Y either has an R-maximal element or a subclass which is a nest and a proper class.

Suppose the first alternative holds. Let m be an R-maximal element of Y, then $\bigcup m$ can be well-ordered by R. The hypothesis of M 2S implies that $\bigcup m$ has an R-upper bound, b. b is an R-maximal element of X.

[1] Maximal with respect to inclusion, $\subseteq$.

If the second alternative holds then there exists a nest $N \subseteq Y$ such that N is a proper class. $\bigcup\bigcup N$ is then a subclass of X which is a proper class and is well-ordered by R, q. e. d.

It is clear that MR 14S(K) $\rightarrow$ M 6S. We shall show next that M 6S $\rightarrow$ M 3S, M 1S $\rightarrow$ M 6S, M 3S $\leftrightarrow$ M 23S, and M 5S $\leftrightarrow$ MR 14S(K).

THEOREM 3.3S: M 6S $\rightarrow$ M 3S.

PROOF: Let X be an arbitrary class. By M 6S, there exists a maximal subclass $N \subseteq X$ which is a nest. If N is a proper class then X has a subclass which is a proper class and is a nest. If N is a set, then by the hypothesis of M 3S, $\bigcup N \in X$ and $\bigcup N$ is clearly a maximal element of X.

THEOREM 3.4S: M 1S $\rightarrow$ M 6S.

PROOF: Let X be an arbitrary class. Let

$$Y = \{x: x \subseteq X \text{ and } x \text{ is a maximal nest in } X \cap \mathscr{P}(\bigcup x)\}.$$

Y is partially ordered by inclusion. Let s be a linearly ordered subset of Y. We note that M 1S $\rightarrow$ M 1 $\rightarrow$ M$'$ 6. M$'$ 6 implies that there is a maximal nest y such that

(i) $\bigcup s \subseteq y \subseteq \mathscr{P}(\bigcup\bigcup s)$.

It is clear that y is an upper bound of s since $\bigcup s \subseteq y$. We claim also that $y \in Y$ because since $\mathscr{P}$ and $\bigcup$ are monotone (i) implies

(ii) $\mathscr{P}(\bigcup\bigcup s) \subseteq \mathscr{P}(\bigcup y) \subseteq \mathscr{P}(\bigcup \mathscr{P}(\bigcup\bigcup s))$.

But for any set z, $\bigcup \mathscr{P}(z) = z$. Therefore, from (ii) we obtain $\mathscr{P}(\bigcup\bigcup s) = \mathscr{P}(\bigcup y)$. Hence, $y \in Y$.

M 1S implies that either Y has a maximal element, which is clearly a maximal nest in X, or Y contains a subclass S which is a proper class and is linearly ordered. $\bigcup S$ is clearly a nest and we claim that $\bigcup S$ is a maximal nest in X. For suppose not. Suppose there exists a $u \in X$ such that $\bigcup S \cup \{u\}$ is a nest. Then either $v \subseteq u$ for all $v \in \bigcup S$ or there exists a $v \in \bigcup S$ such that $u \subset v$. In the first case, $\bigcup\bigcup S \subseteq u$ which is a contradiction since u is a set and S is a proper class. In the second case, $v \in \bigcup S$ implies there exists an $x \in S$ such that $v \in x$. $x \cup \{u\} \subseteq \bigcup S \cup \{u\}$, therefore, $x \cup \{u\}$ is a nest. Moreover, since $u \subset v \in x$, $x \cup \{u\}$

$\subseteq X \cap \mathscr{P}(\bigcup x)$. But this is a contradiction because $x \in Y$, therefore x is supposed to be a maximal nest in $X \cap \mathscr{P}(\bigcup x)$, q. e. d.

THEOREM 3.5S: M 23S $\rightarrow$ M 3S.

PROOF: Let X be an arbitrary class. We define a relation R on $X \times On$ as follows: If $\langle s, \alpha\rangle$ and $\langle t, \beta\rangle \in X \times On$ then $\langle s, \alpha\rangle \, R \, \langle t, \beta\rangle \leftrightarrow s \subset t$ and $\alpha < \beta$ or $s = t$, s is maximal, and $\alpha < \beta$. We define a function F on $X \times On$ as follows:

$$F(\langle s, \alpha\rangle) = \{\{\langle s, \alpha\rangle\}\} \cup \{F(\langle s', \alpha'\rangle) : \langle s', \alpha'\rangle \, R \, \langle s, \alpha\rangle \text{ and } \langle s', \alpha'\rangle \in X \times On\}.$$

It is easily shown that

$$F(\langle s, \alpha\rangle) = F(\langle t, \beta\rangle) \leftrightarrow \langle s, \alpha\rangle = \langle t, \beta\rangle,$$

$$F(\langle s, \alpha\rangle) \in F(\langle t, \beta\rangle) \leftrightarrow \langle s, \alpha\rangle \, R \, \langle t, \beta\rangle \text{ (see 4.60).}$$

M 23S implies that $\{F(\langle s, \alpha\rangle) : \langle s, \alpha\rangle \in X \times On\}$ has a maximal subclass Q such that for every $S, T \in Q$, $S \in T$ or $T \in S$. Therefore $X \times On$ has a maximal subclass K which is linearly ordered by R.

Suppose X has no maximal element. $\mathscr{D}(K)$ is a nest. If $\mathscr{D}(K)$ is a set then by the hypothesis of M 3S, $\bigcup \mathscr{D}(K) \in X$ and therefore, $\bigcup \mathscr{D}(K)$ would be a maximal element of X. Hence, if X has no maximal element $\mathscr{D}(K)$ is a sub-class of X which is a nest and a proper class.

THEOREM 3.6S: M 3S $\rightarrow$ M 23S.

PROOF: Let X be an arbitrary class. For any set x define $N(x) = x \cup \bigcup x \cup \bigcup\bigcup x \cup \ldots$. Also, define $Y = \{y : y \subseteq X$ and y is a maximal subset of $X \cap N(y)$ which is connected by $\in\}$. It follows from M 23 that $Y \neq \Lambda$. If Y has a maximal element that proves M 23S. If Y has no maximal element then M 3S implies that there exists a nest Z contained in Y which is a proper class. We claim $\bigcup Z$ is a maximal subclass of X which is connected by $\in$.

It is clear that $\bigcup Z$ is a subset of X which is connected by $\in$. Suppose $\bigcup Z$ is not maximal. Suppose there is an $x \in X \sim \bigcup Z$ such that $(\{x\} \cup \bigcup Z)$ is connected by $\in$. Therefore, for all y, if $y \in \bigcup Z$ then either $y \in x$ or $x \in y$. Since $\bigcup Z$ is a proper class $\bigcup Z \nsubseteq x$, therefore there exists a $y \in \bigcup Z$ such that $x \in y$. Since $y \in \bigcup Z$ there is a $t \in Z$ such that $y \in t$. Since $Z \subseteq Y$, $t \in Y$, and since $x \in y \in t$, $x \in N(t)$. Since $\{x\} \cup \bigcup Z$ is connected by $\in$, and $t \subseteq \bigcup Z$ it follows that $\{x\} \cup t$ is $\in$-connected. However, t is a maximal subset of $X \cap N(t)$ which is $\in$-connected, therefore, $x \in t$. Hence, $x \in \bigcup Z$. But this is a contradiction so that $\bigcup Z$ is the required class.

THEOREM 3.7S: M 5S $\rightarrow$ MR 14S(K).

PROOF: Let R be an arbitrary relation. Define a relation S as follows:

$$x \, S \, y \leftrightarrow x, y \in \mathscr{D}(R) \text{ and } R''\{x\} \subseteq R''\{y\}.$$

S is a transitive relation so that by M 5S, there exists a maximal subclass X of $\mathscr{D}(R)$ which is linearly ordered by S. X is the required class.

THEOREM 3.8S: MR 14S(K) $\rightarrow$ M 5S.

PROOF: Let R be an arbitrary transitive relation. (We may assume without loss of generality that R is reflexive, for if not just consider $R \cup I$ instead of R). Then we claim

(1) $x \, R \, y \leftrightarrow R''\{y\} \subseteq R''\{x\}$.

For suppose $x \, R \, y$ and $z \in R''\{y\}$. Then $y \, R \, z$ so that by the transitivity of R we obtain $x \, R \, z$ which implies $z \in R''\{x\}$. Now suppose $R''\{y\} \subseteq R''\{x\}$. Since R is reflexive, $y \in R''\{y\}$ which implies $y \in R''\{x\}$ which implies $x \, R \, y$.

Hence, it follows from (1) that MR 14S(K) $\rightarrow$ M 5S.

Next we shall prove:

(II) WE 5S $\rightarrow$ MF 3S $\leftrightarrow$ AL′ 3S $\leftrightarrow$ M 7S $\rightarrow$ M 13S $\rightarrow$ M 8S $\leftrightarrow$ M 9S $\leftrightarrow$ M 10S $\leftrightarrow$ M 11S $\leftrightarrow$ M 12S $\leftrightarrow$ MR 14S(A) [1] $\rightarrow$ M 5S.

The arguments that the following chain of implications hold are analogous to the corresponding arguments for the set forms.

M 7S $\rightarrow$ M 13S $\rightarrow$ M 8S $\rightarrow$ M 9S $\leftrightarrow$ M 10S $\rightarrow$ M 12S $\leftrightarrow$ M 11S $\rightarrow$ M 5S.

We can also prove the following without use of the axiom of regularity:

THEOREM 3.9S: M 11S $\rightarrow$ M 8S.

[1] Where $A = D, J, \bar{J}, U, \bar{U}, DK, DJ, DU, KJ, KU$, or JU.

PROOF: Let R be an arbitrary relation. We define a relation $*R$ as follows:

$$\langle x, y\rangle \in *R \leftrightarrow \{x, y\} \times \{x, y\} \subseteq R.$$

For any class X, $X \times X \subseteq *R \leftrightarrow X \times X \subseteq R$. Since, $*R \cup *R^{-1} \cup I = *R$, M 11S $\rightarrow$ M 8S.

THEOREM 3.10S: M 11S $\rightarrow$ MR 14S(D).

PROOF: Let R be an arbitrary relation. Define a relation S as follows:

$$x\ S\ y \leftrightarrow x, y \in \mathscr{D}(R),\ x \neq y, \text{ and } R''\{x\}\ D\ R''\{y\}.$$

$S = S^{-1}$, therefore, M 11S implies that there exists a maximal subclass Y of X such that $Y \times Y \subseteq S \cup I$. This is an equivalent way of saying that Y is a maximal subclass of X such that for all $x, y \in Y$, $x \neq y$, $R''\{x\}\ D\ R''\{y\}$.

In an analogous way we can prove that M 11S $\rightarrow$ MR 14S(A) if $A = \bar{D}, J, \bar{J}, U, \bar{U}, DK, DJ, DU, KJ, KU$, or JU. In a similar way we can prove M 13S $\rightarrow$ MR 14S(A) if $A = D_m$ or $\bar{D}_m$.

THEOREM 3.11S [1]: MR 14S(D) $\rightarrow$ M 11S.

PROOF: Let R be an arbitrary relation. Define a relation S as follows: If $x \in \mathscr{D}(R)$,

$$x\ S\ y \leftrightarrow (\exists z)[y = \{x, z\} \text{ and } \langle x, z\rangle \in (\bar{R} \cap \bar{R}^{-1}) \cup I].$$

(That is, $\langle x, z\rangle \notin R$ and $\langle z, x\rangle \notin R$ or $x = z$.)

If $x, y \in \mathscr{D}(S)$, $x \neq y$, then $S''\{x\} \nsubseteq S''\{y\}$ because $\{x\} \in S''\{x\}$ but $\{x\} \notin S''\{y\}$. If $x \neq y$, then $S''\{x\} \cap S''\{y\} \subseteq \{\{x, y\}\}$.

If $X \subseteq \mathscr{D}(R) = \mathscr{D}(S)$, the following statements are all equivalent:

(i) $S''\{x\}\ D\ S''\{y\}$, for all $x, y \in X$, $x \neq y$.

(ii) $S''\{x\} \cap S''\{y\} = \Lambda$, for all $x, y \in X$, $x \neq y$.

(iii) $\langle x, y\rangle \notin \bar{R} \cap \bar{R}^{-1}$, for all $x, y \in X$, $x \neq y$.

(iv) $\langle x, y\rangle \in R \cup R^{-1}$, for all $x, y \in X$, $x \neq y$.

(v) $X \times X \subseteq R \cup R^{-1} \cup I$, q. e. d.

[1] We will prove that the relations in the preceding footnote are *universal*, that is, given any symmetric relation R and any of the above A, there is a relation S such that whenever $x \neq y$, xRy if and only if $S''\{x\}A S''\{y\}$. A similar remark holds for the set forms and was used by Kurepa [1]. For D and $\bar{D}$ this was previously observed by Szpilrajn-Marczewski [1].

The same proof will hold if D is replaced by DK, DU or $\bar{J}$. By duality an analogous proof will hold for U and KU.

THEOREM 3.12S: MR 14S($\bar{D}$) $\rightarrow$ M 11S.

PROOF: Let R be an arbitrary relation. Define a relation S as follows: If $x \in \mathscr{D}(R)$

$$x \; S \; y \leftrightarrow (\exists z)[y = \{x, z\} \text{ and } \langle x, z\rangle \in R \cup R^{-1} \cup I].$$

Just as in 3.11S we have if $x, y \in \mathscr{D}(S)$, $x \neq y$, then $S''\{x\} \nsubseteq S''\{y\}$ and $S''\{x\} \cap S''\{y\} \subseteq \{\{x, y\}\}$.

If $X \subseteq \mathscr{D}(R) = \mathscr{D}(S)$, the following statements are all equivalent:

(i) $S''\{x\} \, \bar{D} \, S''\{y\}$, for all $x, y \in X$, $x \neq y$.

(ii) $S''\{x\} \cap S''\{y\} \neq \Lambda$, for all $x, y \in X$, $x \neq y$.

(iii) $\langle x, y\rangle \in R \cup R^{-1}$, for all $x, y \in X$, $x \neq y$.

(iv) $X \times X \subseteq R \cup R^{-1} \cup I$, q. e. d.

The same proof will hold if $\bar{D}$ is replaced by J, JU or JK. By duality an analogous proof will hold for $\bar{U}$ and DJ.

THEOREM 3.13S: AL′ 3S $\rightarrow$ M 7S.

PROOF: Let the closure operator in AL′ 3S be the identity operator; then clearly AL′ 3S $\rightarrow$ M 7S.

THEOREM 3.14S: MF 3S $\rightarrow$ AL′ 3S.

PROOF: Let X be an arbitrary class, P a property of finite character, and C' a finitary closure operator. Let $C(X) = C'(X) \cup X$, then X is C-closed if and only if X is C'-closed and $X \subseteq C(X)$. Let Y be a C-closed subclass of X. Let C^* be defined as follows:

$$C^*(Z) = \bigcup_{n=0}^{\infty} C^n(Z).$$

$C^*(Z)$ is the smallest C-closed class containing Z. C^* is clearly a finitary closure operator. Define a class-valued function F as follows:

$$\mathscr{D}(F) = \{x\colon x \subseteq X \text{ and } P[C^*(x \cup Y)]\}.$$

$$x \in \mathscr{D}(F) \rightarrow F[x] = C^*(x \cup Y).$$

If $x \subseteq Y$, $F[x] = Y$. Define a predicate Π as follows:

$$\Pi(Z) \leftrightarrow Z = C(Z) \text{ and } P[Z] \text{ and } Y \subseteq Z \subseteq X.$$

Now we must show that the hypotheses of MF 3S are satisfied. It follows from the definition of Π, $\mathscr{D}(F)$ and F that $\Pi(Z) \leftrightarrow (x)[x \subseteq Z \rightarrow x \in \mathscr{D}(F)$ and $F[x] \subseteq Z]$. Suppose Z is a non-empty nest contained in $\mathscr{D}(F)$. Consider $W = \bigcup_{x \in Z} C^*(x \cup Y)$. Suppose $v \in C(W)$, then there exists a finite subset y of W such that $v \in C(y)$. Since Z is a nest, there exists a $z \in Z$ such that $x \in C(C^*(z \cup Y)) = C^*(z \cup Y) \subseteq W$. Therefore $C(W) \subseteq W$, which implies $W = C(W)$. It follows from the definition of $\mathscr{D}(F)$ that W has the property P and it is clear that $Y \subseteq W \subseteq X$. Hence $\Pi(W)$ and the hypotheses of MF 3S are satisfied. Therefore, there exists a maximal class which has the property Π, q. e. d.

Before preceding with the proof that M 7S $\rightarrow$ MF 3S we prove the following theorem:

THEOREM 3.15S: M 11S $\rightarrow$ AC 4S.

PROOF: Let R be an arbitrary relation. Define a relation S as follows:

$$\langle x, y\rangle \, S \, \langle z, w\rangle \leftrightarrow x \neq z.$$

M 11S implies that there exists a maximal class F contained in R such that $F \times F \subseteq S \cup S^{-1} \cup I$. F is the required function.

THEOREM 3.16S: M 7S $\rightarrow$ MF 3S.

PROOF: Let F be a class-valued function and Π a predicate which satisfy the hypothesis of MF 3S. That is,

(i) $\Pi[x] \leftrightarrow (x)[x \subseteq X \rightarrow x \in \mathscr{D}(F)$ and $F[x] \subseteq X]$

(ii) $x \subseteq F[x]$ for all $x \in \mathscr{D}(F)$

(iii) If for every non-empty nest $Y \subseteq \mathscr{D}(F)$, $\Pi[\bigcup_{y \in Y} F[y]]$.

It follows from (iii) that

(1) $x \in \mathscr{D}(F) \rightarrow \Pi[F[x]]$ (take Y to be $\{x\}$).

Let

(2) $P = \{x: x \in \mathscr{D}(F)\}$.

We wish to show that P is a property of finite character. Suppose $P[x]$ and $y \subseteq x$. Therefore, $x \in \mathscr{D}(F)$ by (2); $y \in \Pi[F[x]]$ by (1); $x \subseteq F[x]$ by (ii); $y \subseteq F[x]$ since $y \subseteq x$; $y \in \mathscr{D}(F)$ by (i); and $P[y]$ by (2). Now suppose for all $y \subseteq x$ if y is finite then $P[y]$. We wish to prove $P[x]$. Since M 7S $\rightarrow$ M 7 $\rightarrow$ WE 1, $\overline{\overline{x}}$ is either finite or an aleph. Clearly if $\overline{\overline{x}}$ is finite then $P[x]$. Suppose the theorem is false, then there exists a smallest ordinal number α such that there is an x with $\overline{\overline{x}} = \aleph_\alpha$ and for all $y \subseteq x$ if y is finite then $P[y]$. Let ω_α be the smallest ordinal number such that $\overline{\overline{\omega_\alpha}} = \aleph_\alpha$, then $\omega_\alpha \approx x$. Let f be a 1-1 function mapping ω_α onto x. Let $z_\beta = f''\beta$ for $\beta \in \omega_\alpha$. If $\beta < \omega_\alpha$ then $\overline{\overline{z_\beta}} < \aleph_\alpha$, which implies $z_\beta \in \mathscr{D}(F)$. Since $\{z_\beta \colon \beta < \omega_\alpha\}$ forms a nest, it follows from (iii) that $\Pi[\bigcup_{\beta<\omega_\alpha} F[z_\beta]]$. Since $x = \bigcup_{\beta<\omega_\alpha} z_\beta \subseteq \bigcup_{\beta<\omega_\alpha} F[z_\beta]$ by (ii), it follows from (i) that $x \in \mathscr{D}(F)$. Therefore $P[x]$ by (2). But this is a contradiction. Hence, P is a property of finite character.

Next, we define a property Q as follows:

(3) $Q[X] \leftrightarrow (\exists Y)[X \subseteq Y \text{ and } \Pi[Y]]$.

If we can show that Q is a property of finite character then this will complete the proof of the theorem for M 7S then implies that there exists a maximal class with the property Q and this class is also a maximal class with the property Π. In order to prove that Q is a property of finite character we shall prove

(4) $Q[X] \leftrightarrow (y)[y \subseteq X \rightarrow P[y]]$.

Suppose $Q[X]$ holds, then there is a Y such that $X \subseteq Y$ and $\Pi[Y]$. If $y \subseteq X$ then $y \subseteq Y$. Therefore, by (i), $y \in \mathscr{D}(F)$. Hence, by (2), $P[y]$.

Now suppose

(5) $(y)[y \subseteq X \rightarrow P[y]]$.

We wish to prove $Q[X]$. Let

(6) $Y = \bigcup_{y \subseteq X} F[y]$.

It follows from (5) and (2) that for all $y \subseteq X$, $y \in \mathscr{D}(F)$, and it follows from (ii) that $X \subseteq Y$. In order to prove $Q[X]$ it is sufficient to prove $\Pi[Y]$.

It follows from the remarks preceding 3.9S that M 7S $\to$ M 11S. Hence, by 3.15S we may assume that AC 4S holds. We define a relation R as follows: $R = \{\langle u, y\rangle : u \in F[y] \text{ and } y \subseteq X\}$. Then AC 4S implies that there exists a function G such that $\mathscr{D}(G) = \mathscr{D}(R)$ and $G \subseteq R$. Let $z \subseteq Y$ and let $y = \bigcup G''z = \bigcup_{u \in z} G(u)$. By the definition of G, $y \subseteq X$, therefore $y \in \mathscr{D}(F)$ by (5) and (2). Therefore (1) implies $\Pi[F[y]]$. Suppose $u \in z$, then since $z \subseteq Y$, $u \in F[G(u)] \subseteq F[y]$. Therefore, $z \subseteq F[y]$ so that $\Pi[F[y]]$ and (i) imply

(7) $z \in \mathscr{D}(F)$

and

(8) $F[z] \subseteq F[y]$.

But $F[y] \subseteq Y$. Hence

(9) $F[z] \subseteq Y$.

Therefore, if $z \subseteq Y$, (7) and (9) hold, so that (i) implies $\Pi[Y]$, which implies $Q[X]$, q. e. d.

We shall prove next that WE 5S $\to$ MF 3S.

THEOREM 3.17S: WE 5S $\to$ MF 3S.

PROOF: Let F be a class-valued function and Π a predicate which satisfy the hypothesis of MF 3S. WE 5S implies that there exists a 1-1 function G which maps On onto V. Define another function H such that for each ordinal number α,

$$H(\alpha) = \begin{cases} \bigcup_{\beta<\alpha} H(\beta) \cup \{G(\alpha)\} \text{ if } \bigcup_{\beta<\alpha} H(\beta) \cup \{G(\alpha)\} \in \mathscr{D}(F), \\ \bigcup_{\beta<\alpha} H(\beta) \text{ otherwise.} \end{cases}$$

By the definition of H it follows that if $\alpha < \beta$ then $H(\alpha) \subseteq H(\beta)$. Therefore, $H''On$ is a nest. We shall also show that $H(\alpha) \in \mathscr{D}(F)$ for every $\alpha \in On$. By the hypothesis of MF 3S, there exists a class which has the property Π. Therefore, $\Lambda \in \mathscr{D}(F)$ so that $H(0) \in \mathscr{D}(F)$. Now, suppose for $\beta < \alpha$, $H(\beta) \in \mathscr{D}(F)$ which implies $\bigcup_{\beta<\alpha} H(\beta) \subseteq \mathscr{D}(F)$. Since $\bigcup_{\beta<\alpha} H(\beta)$ is a nest, by the hypothesis of MF 3S, $\Pi(\bigcup_{\beta<\alpha} F[H(\beta)])$. Again by the hypothesis of MF 3S, we have $H(\beta) \subseteq F[H(\beta)]$ for all $\beta < \alpha$, which implies $\bigcup_{\beta<\alpha} H(\beta) \subseteq \bigcup_{\beta<\alpha} F[H(\beta)]$. Hence, by the definition of Π, we have $\bigcup_{\beta<\alpha} H(\beta) \in \mathscr{D}(F)$ and this implies $H(\alpha) \in \mathscr{D}(F)$.

We have obtained now that $H''On$ is a nest contained in $\mathscr{D}(F)$, hence, by the hypothesis of MF 3S, $\Pi(\bigcup_{\alpha \in On} F[H(\alpha)])$. Let $A = \bigcup_{\alpha \in On} F[H(\alpha)]$. We claim that A is a maximal class with the property Π, for suppose $A \subseteq B$ and $\Pi(B)$. Since G maps On onto V, let $G(\alpha)$ be an arbitrary element of B. We have $\bigcup_{\beta<\alpha} H(\beta) \subseteq A$, so that $\bigcup_{\beta<\alpha} H(\beta) \cup \{G(\alpha)\} \subseteq B$. Therefore, $\Pi(B)$ implies $\bigcup_{\beta<\alpha} H(\beta) \cup \{G(\alpha)\} \in \mathscr{D}(F)$. Hence, $G(\alpha) \in H(\alpha) \subseteq F[H(\alpha)] \subseteq A$. This implies that $B = A$, q. e. d.

We shall show next,

(III) MR 14S($\bar{J}$) $\to$ M 14S($\bar{J}$) $\to$ M 14S(D) $\to$ AC 2S.

It is clear that MR 14S($\bar{J}$) $\to$ M 14S($\bar{J}$).

THEOREM 3.18S: M 14S($\bar{J}$) $\to$ M 14S(D).

PROOF: Let X be an arbitrary class. For each $s \in X$, define $s_u = s \cup \{\langle s, u\rangle\}$ where $(t)[t \in X$ and $t \neq s \to \langle s, u\rangle \notin t]$. If $s \neq t$, $s_u \nsubseteq t_u$ which implies $s_u \, D \, t_u \leftrightarrow s_u \, \bar{J} \, t_u$. If $s \neq t$, $s_u \cap t_u = s \cap t$, so that $s_u \, D \, t_u \leftrightarrow s \, D \, t$. M 14S($\bar{J}$) implies that $\{s_u : s \in X\}$ has a maximal subclass with the property $\bar{J}$, hence X has a maximal subclass with the property D. (See 4.16.)

THEOREM 3.19S: M 14S(D) $\to$ AC 2S.

PROOF: Analogous to 4.17.

Some of the propositions which have been stated contain the condition that the universe can be linearly ordered. (M 14S(J) and ($\bar{K}$) and MR 14S($\bar{K}$)). Without using the axiom of regularity, the weakest form of the axiom of choice which is known to imply that the universe can be linearly ordered is M 7S. We shall demonstrate this proof in the following lemma.

Let us introduce the following notation:

LV: *V can be anti-symmetrically linearly ordered.*

LEMMA 3.20S: M 7S $\to$ LV.

PROOF: Let X be a class of ordered pairs. We say X is *circular* if there exists a finite sequence of length $n > 1$ of ordered pairs in X, $\langle z_0, z_1\rangle$, $\langle z_1, z_2\rangle$, $\langle z_2, z_3\rangle$, ..., $\langle z_{n-1}, z_n\rangle$, such that $z_0 = z_n$, and the other z_i's are distinct. If no such sequence exists X is called *non-circular*. Let P be a property defined as follows: For every class of ordered pairs A,

$$P[A] \leftrightarrow A \text{ is non-circular.}$$

P is clearly a property of finite character.

M 7S implies that there exists a maximal class, R, with the property P. We claim R is an anti-symmetric linear ordering of the universe. R is clearly anti-symmetric. Suppose $\langle x, y\rangle$ and $\langle y, z\rangle \in R$ and $\langle x, z\rangle \notin R$. This implies $R \cup \{\langle x, z\rangle\}$ is circular. However, if $R \cup \{\langle x, z\rangle\}$ is circular then clearly R is circular which is a contradiction. Therefore, $\langle x, z\rangle \in R$ and R is transitive. Suppose $\langle x, y\rangle \notin R$ then $R \cup \{\langle x, y\rangle\}$ is circular. So that there exists a finite sequence of ordered pairs $\langle z_0, z_1\rangle$, $\langle z_1, z_2\rangle$, ..., $\langle z_{n-1}, z_n\rangle$ in $R \cup \{\langle x, y\rangle\}$ such that $z_0 = z_n$ and since R is not circular one of the ordered pairs is $\langle x, y\rangle$ say $\langle z_i, z_{i+1}\rangle$. Each of the following are elements of R: $\langle z_{i+1}, z_{i+2}\rangle$, $\langle z_{i+2}, z_{i+3}\rangle$, ..., $\langle z_{n-1}, z_n\rangle$, $\langle z_0, z_1\rangle$, ..., $\langle z_{i-1}, z_i\rangle$, where $z_{i+1} = y$ and $z_i = x$. Therefore, since R is transitive it follows that $\langle y, x\rangle \in R$. Hence, R is connected. Since R is transitive and connected it is a linear ordering of the universe.

We shall now proceed to prove the remaining propositions of this section equivalent.

THEOREM 3.21S: LV *and* M 11S $\rightarrow$ MR 15S($\bar{K}$).

PROOF: Substitute $\bar{K}$ for D in 3.10S.

THEOREM 3.22S: MR 15S($\bar{K}$) $\rightarrow$ AC 4S.

PROOF: Let R be an arbitrary relation and let L be a reflexive relation which anti-symmetrically linearly orders the universe. Define a relation S as follows:

$$S = \{\langle\langle x, y\rangle, \langle x, z\rangle\rangle : \langle x, y\rangle \in R \text{ and } \langle x, z\rangle \in R \text{ and } \langle y, z\rangle \in L\}.$$

If $\langle x, y\rangle$ and $\langle z, w\rangle \in \mathscr{D}(S)$ then $S''\{\langle x, y\rangle\}\, K\, S''\{\langle z, w\rangle\} \leftrightarrow x = z$. For, first of all, if $x \neq z$ then $S''\{\langle x, y\rangle\} \cap S''\{\langle z, w\rangle\} = \Lambda$. Now suppose $x = z$. Either $\langle y, w\rangle \in L$ or $\langle w, y\rangle \in L$. Suppose $\langle y, w\rangle \in L$ and suppose $\langle u, v\rangle \in S''\{\langle z, w\rangle\}$. Then $u = z$ and $\langle w, v\rangle \in L$. Since $\langle y, w\rangle$ and $\langle w, v\rangle \in L$, $\langle y, v\rangle \in L$. Hence $\langle u, v\rangle \in S''\{\langle x, y\rangle\}$ which implies $S''\{\langle z, w\rangle\} \subseteq S''\{\langle x, y\rangle\}$. If $\langle w, y\rangle \in L$ we obtain in a similar manner, $S''\{\langle x, y\rangle\} \subseteq S''\{\langle z, w\rangle\}$.

MR 15S($\bar{K}$) implies there exists a maximal class F contained in $\mathscr{D}(S)$ such that if $s, t \in F$, $s \neq t$, then $S''\{s\} \bar{K} S''\{t\}$. Clearly $\mathscr{D}(F) = \mathscr{D}(R)$ and $F \subseteq R$. If $\langle x, y\rangle$ and $\langle x, z\rangle \in \mathscr{D}(S)$ then $S''\{\langle x, y\rangle\} K S''\{\langle x, z\rangle\}$, hence not both $\langle x, y\rangle$ and $\langle x, z\rangle$ belong to F. Therefore F is a function which satisfies AC 4S. (This proof is similar to 4.19.)

It is clear that if the universe can be linearly ordered and MR 14S(J) holds then M 15S(J) also holds.

THEOREM 3.23S: M 15S(J) $\rightarrow$ M 15S($\bar{K}$).

PROOF: Substitute J for $\bar{J}$ and $\bar{K}$ for K in 4.18.

THEOREM 3.24S: M 15S($\bar{K}$) $\rightarrow$ AC 2S.

PROOF: Analogous to 4.19.

THEOREM 3.25S: AC 4S $\rightarrow$ M 14S(K) (= M 6S).

PROOF: Let X be an arbitrary class of sets. Let R be a relation defined as follows:

$$R = \{\langle x, s\rangle : x \subseteq X \text{ and } s \in X \text{ and } (t)[t \in x \rightarrow t \subset s]\}.$$

By AC 4S, there exists a function F such that $\mathscr{D}(F) = \mathscr{D}(R)$ and $F \subseteq R$.

Suppose $n \subseteq X$ and n is a nest which has no strict upper bound. Let m be a nest such that $n \subseteq m \subseteq X$. Suppose $s \in m \sim n$. Then, since n has no strict upper bound, there exists a $t \in n$ such that $s \subseteq t$. This implies $s \subseteq \bigcup n$. Hence, $m \subseteq \mathscr{P} \bigcup n \cap X$. Since $\mathscr{P} \bigcup n$ is a set, this implies that M′ 6 $\rightarrow$ M 14S(K) in this case. AC 4S clearly implies AC 4 and it was shown in Part I that AC 4 $\rightarrow$ M′ 6. In the remaining part of the proof we shall assume that every nest has a strict upper bound.

Define a relation S as follows: For all x, y, $\langle x, y\rangle S z$ if z is a maximal nest such that $x \subseteq z \subseteq \mathscr{P}(y) \cap X$. M′ 6 implies the existence of such a z whenever x is a nest and y is a set such that $x \subseteq \mathscr{P}(y) \cap X$. AC 4S implies that there exists a function H

such that $\mathscr{D}(H) = \mathscr{D}(S)$ and $H \subseteq S$. We define a function G for each $\alpha \in On$ as follows:

$$G(\alpha) = H(\bigcup G''\alpha, F(\bigcup G''\alpha)).$$

We are assuming here that every nest has a strict upper bound. Therefore, in order to prove that $G(\alpha)$ exists for every $\alpha \in On$ it is sufficient to prove that $\bigcup G''\alpha$ is a nest for every $\alpha \in On$. If $\alpha = 0$, $\bigcup G''0 = \Lambda$ which is a nest. Suppose $\bigcup G''\alpha$ is a nest for every $\alpha < \beta$.

Case 1. β is a limit ordinal, Let $x, y \in \bigcup G''\beta$, then there exist ordinal numbers γ and δ, $\gamma < \beta$, $\delta < \beta$, such that $x \in G(\gamma)$ and $y \in G(\delta)$. Since β is a limit ordinal there exists an ordinal number $\alpha < \beta$ such that $G(\gamma) \in G''\alpha$ and $G(\delta) \in G''\alpha$. Hence, $x, y \in \bigcup G''\alpha$ and by induction hypotheses $\bigcup G''\alpha$ is a nest. Therefore, $x\,K\,y$.

Case 2. β is not a limit ordinal. Then there exists an ordinal number α such that $\beta = \alpha + 1$. $G''\beta = G''(\alpha + 1) = G''\alpha \cup \{G(\alpha)\}$, so that $\bigcup G''\beta = \bigcup G''\alpha \cup G(\alpha)$. $\bigcup G''\alpha$ is a nest by the induction hypothesis, therefore, by the definition of H, $G(\alpha)$ is a nest and $\bigcup G''\alpha \subseteq G(\alpha)$. Hence, $G''\beta = G(\alpha)$ is a nest.

Next, we shall prove that $\bigcup G''On$ is a maximal nest contained in X. It is clear that $\bigcup G''On \subseteq X$ and that it is a nest since whenever $\alpha, \beta \in On$ and $\alpha < \beta$ then $G''\alpha \subseteq G''\beta$. Suppose there exists an $x \in X$ such that $\bigcup G''On \cup \{x\}$ is a nest. Since $G''On$ is a proper class, we cannot have that for all $u \in \bigcup G''On$, $u \subseteq x$. Therefore, there exists a $u \in \bigcup G''On$ such that $x \subseteq u$. So that there is an $\alpha \in On$ such that $u \in G''\alpha$. Hence, by the maximality of $G(\alpha)$ we have $x \in G(\alpha) \subseteq \bigcup G''On$, q. e. d.

The following implications have now been proved:

(IV) LV and M 11S $\to$ MR 15S($\bar{K}$) $\to$ AC 4S $\to$ M 14S(K).
LV and MR 14S(J) $\to$ M 15S(J) $\to$ M 15S($\bar{K}$) $\to$ AC 2S.

Next we shall prove: For all natural numbers $m \geq 2$.

$$\text{(V)}\quad \begin{array}{ccccccc}
\text{MR 14S}(\bar{D}_m) & \leftrightarrow & \text{MR 14S}(D_m) & \to & \text{M 14S}(D_m) & \to & \text{AC } 7(m-1)\text{S} \\
\downarrow & & \downarrow & & \downarrow & & \downarrow \\
\text{MR 18S} & \leftrightarrow & \text{MR 16S} & \to & \text{M 16S} & \to & \text{AC 8S} \\
\downarrow & & \downarrow & & \downarrow & & \downarrow \\
\text{MR 19S} & \to & \text{MR 17S} & \to & \text{M 17S} & \to & \text{AC 9S}
\end{array}$$

The verticle implications hold for purely logical reasons and it is also clear that MR 14S(D_m) $\to$ M 14S(D_m), MR 16S $\to$ M 16S, and MR 17S $\to$ M 17S. The proofs of the following implications are similar to the proofs for the corresponding set forms: MR 14S($\bar{D}_m$) $\to$ MR 14S(D_m), MR 18S $\to$ MR 16S, MR 19S $\to$ MR 17S (See 4.23); M 14S(D_m) $\to$ AC 7($m - 1$)S, M 16S $\to$ AC 8S, M 17S $\to$ AC 9S (See 4.21). It is not known whether it is possible to prove the following equivalences without making use of the axiom of regularity: MR 17S $\to$ MR 19S, and if $m \neq n$, MR 14S($\bar{D}_m$) $\to$ MR 14S($\bar{D}_n$), MR 14S(D_m) $\to$ MR 14S(D_n), M 14S(D_m) $\to$ M 14S(D_n). In order to prove the two remaining implications in (V) (MR 14S(D_m) $\to$ MR 14S($\bar{D}_m$) and MR 16S $\to$ MR 18S) we shall prove a lemma.

DEFINITION 3.26S: If P is a property of m classes, R is a relation and X is a class then

$$P_R[X] \text{ if and only if } P[R''\{x_1\}, R''\{x_2\}, \ldots, R''\{x_m\}]$$

for all $x_1, x_2, \ldots, x_m \in X$ such that the x_i's are distinct.

Using this notation we could have written MR 14S(A) as follows:

For every relation R and class X there exists a maximal subclass Y such that $A_R[Y]$.

LEMMA 3.27S: *For every class relation R and for every natural number $m \geq 2$, there exists a relation S such that $\bar{D}_{mR}[X]$ if and only if $D_{mS}[X]$ for every class X.*

PROOF: Let R be a relation, m a natural number, $m \geq 2$, and X a class. Define S as follows:

$$S = \{\langle x, y\rangle : x \in \mathscr{D}(R),\ y \subseteq \mathscr{D}(R),\ x \in y, \text{ and } (y = \{x\} \text{ or } (y \approx m \text{ and } D_{mR}[y]))\}.$$

Now we have,

$$\begin{aligned} D_{mS}[X] &\leftrightarrow (y)[y \subseteq X \text{ and } y \approx m \to \bigcap_{x \in y} S''\{x\} = \Lambda] \\ &\leftrightarrow (y)[y \subseteq X \text{ and } y \approx m \to \bar{D}_{mR}[y]] \\ &\leftrightarrow \bar{D}_{mR}[X]. \end{aligned}$$

By means of 3.27S it is easy to show that MR 14S(D_m) $\to$ MR 14S($\bar{D}_m$) and MR 16S $\to$ MR 18S.

All the implications in (I)–(V) were proved without use of the axiom of regularity. In order to prove that all the preceding class forms are equivalent it remains to be shown that one of M 1S–M 4S implies WE 4S or WE 5S.

THEOREM [1] 3.28S: M 1S $\rightarrow$ WE 4S.

PROOF: Define a class A as follows:

$A = \{\langle x, w\rangle : w \subseteq x \times x,\ w$ well-orders x and $(s)(t)[s \in x$ and $\rho(t) < \rho(s) \rightarrow t \in x]\}$. (We assume $\mathscr{I}$ is a set.)
(Note: M 1 S $\rightarrow$ M 1 $\leftrightarrow$ WE 1, therefore A is a proper class.) We define a relation R on A as follows: If $\langle x, w\rangle, \langle x', w'\rangle \in A$ then $\langle x, w\rangle R \langle x', w'\rangle \leftrightarrow x \subseteq x'$ and $w = w' \cap (x \times x)$ and $x \times (x' \sim x) \subseteq w'$. R is clearly a transitive relation on A and if b is an R-linearly ordered subset of A, then $\langle \bigcup \mathscr{D}(b), \bigcup \mathscr{R}(b)\rangle$ is an R-upper bound for b.

M 1S implies that either A has an R-maximal element or A has a subclass which is a proper class which is linearly ordered by R. But A has no R-maximal element. Therefore, let K be a subclass of A which is a proper class and is linearly ordered by R. Let $X = \bigcup \mathscr{D}(K)$ and $W = \bigcup \mathscr{R}(K)$. We shall show that $X = V$ and W well-orders X. Let $s \in V$. Then since X is a proper class there is an $x \in X$, such that $\rho(s) < \rho(x)$. Let $x \in y \in \mathscr{D}(K)$, then by the definition of A, this implies that $s \in y$, which implies that $s \in X$. Since K is linearly ordered by R, it follows from the definition of A and R that W well-orders $X = V$.

Now, we must show that every proper initial W-section of V is a set. Suppose $s \in V$, then there is an $x \in \mathscr{D}(K)$ such that $s \in x$. Suppose $t\,W\,s$ and suppose $t \notin x$, then there exists an $\langle x', w'\rangle \in K$ such that $t \in x'$. Since K is linearly ordered by R, either $\langle x, w\rangle R \langle x', w'\rangle$ or $\langle x', w'\rangle R \langle x, w\rangle$. But since $t \in x'$ and $t \notin x$, we must have $\langle x, w\rangle R \langle x', w'\rangle$. Therefore $\langle s, t\rangle \in x \times (x' \sim x) \subseteq w'$, which implies $s\,W\,t$. But, since $s \neq t$, this contradicts the supposition $t\,W\,s$. Hence $t \in x$ and $\{t : t\,W\,s\} \subseteq x$. Therefore, every proper initial W-section of the universe is a set, q. e. d.

(Note, 4.9 is a similar proof which holds for sets.)

[1] The axiom of regularity was used in the proof.

4. Additional Forms

The forms in this section are class forms for several of the forms in Part I, Section 7. The forms which involve cardinal numbers (Part I, Section 6) can not be strengthened since a proper class does not have a cardinal number.

P 1S is a class form of P 9 and P 10 and is due to von Neumann [1] and [2]. (See the introduction to Part II.)

P 1S: *Every proper class can be mapped onto the universe.*

P 1S is clearly equivalent to WE 5S.

The forms P 2S–P 4S, P 2sS, and P 3sS are class forms of P 13–P 15, P 13s, and P 14s and are obtained by allowing A to be a proper class. However, if A is a proper class then $\bigtimes_{a \in A} B_a = \Lambda$, so that P 2S, P 3S, P 2sS, P 3sS, and P 4S have to be reworded.

P 2S: *For every non-empty class A, for every function B whose domain is A and for each $a \in A$, B_a is a set, and for every function X whose domain is $B \circ \epsilon^{-1}$,*

$$x \in \bigcap_{a \in A} \bigcup_{b \in B_a} X_{a,b} \leftrightarrow (\exists F)[F \text{ is a function}, \mathscr{D}(F) = A, (a)[a \in A \rightarrow F(a) \in B_a], \text{ and } (a)[a \in A \rightarrow x \in X_{a,F(a)}]].$$

Similarly, for P 3S, P 2sS, P 3sS, and P 4S. The proofs that the class distributive laws are equivalent to the other class forms are analogous to the corresponding proofs for the set forms.

P 5S — P 7S are class forms of P 17 and P 18.

P 5S: *For every proper class X, $\mathscr{P}(X) \preccurlyeq X$.*

P 6S: *The power class of a well-ordered class can be well-ordered.*

P 7S: *Every linearly ordered class can be well-ordered.*

Clearly WE 2S → P 7S and WE 5S → P 5S → P 6S. The proofs that P 7S → P 6S and P 6S → WE 4S are analogous to the corresponding proofs for the set forms.

P 8S is metamathematical in form and has no set counterpart. Hilbert [2] in 1923 stated it in the following form:

There exists a function F such that for every property P if $P[F(P)]$ then for all x, $P[x]$.

However, such a statement is meaningless in the system of set theory in which we are working. F is some sort of a metamathematical function which has not been defined. Through discussions with A. Levy and R. Vaught we were able to come up with a more meaningful formulation. To do this we introduce the metamathematical notion of the *iota operator*.

DEFINITION 4.1S: If φ is a propositional function then:

$$X = (\iota Y)\varphi(Y) \leftrightarrow [(\exists 1\ Z)\varphi(Z) \text{ and } \varphi(X)] \text{ or } [\text{not } (\exists 1\ Z)\varphi(Z) \text{ and } X = V].$$

$$((\exists 1\ Z)\varphi(Z) \leftrightarrow (\exists Z)\varphi(Z) \text{ and } (X)(Y)[\varphi(X) \text{ and } \varphi(Y) \rightarrow X = Y].)$$

By introducing the iota operator we avoid the necessity of talking about functions which have proper classes in their domains or ranges. Now, by using this metamathematical notion we are able to rewrite Hilbert's proposition in the following form:

P 8S: *There is a propositional function φ such that*

$$(\exists X)(Y)[Y \neq \Lambda \rightarrow (\iota Z)\varphi(Z, X, Y) \in Y].$$

(Actually, P 8S was obtained by rewriting the contrapositive of Hilbert's original proposition.)

THEOREM 4.2S: WE 2S $\rightarrow$ P 8S.

PROOF: Let $\varphi(Z, X, Y)$ mean Z is the X-first element of Y. WE 2S implies that there exists a relation X which well-orders the universe. Therefore, we have for all non-empty classes Y, $(\iota Z)\varphi(Z, X, Y) \in Y$, q. e. d.

THEOREM 4.3S: P 8S $\rightarrow$ AC 4S.

PROOF: Let R be an arbitrary relation and let φ be a propositional function and X a class which satisfy P 8S. We define a function F as follows:

$$\langle x, y\rangle \in F \leftrightarrow x \in \mathscr{D}(R) \text{ and } y = (\iota Z)\varphi(Z, X, R''\{x\}).$$

Then it follows from P 8S that F is a function which satisfies AC 4S.

Now, in conclusion, we shall summarize the results of Part II. The following implications were all proved without use of the axiom of choice or the axiom of regularity:

A: P 1S ↔ WE 4S ↔ WE 5S ↔ WE 6(m)S ↔ WE 7S ↔ WE 8S.

B: AC 4S ↔ AC 5S ↔ P 2sS ↔ P 3sS.

C: WE 3S ↔ AC 1S ↔ AC 2S ↔ AC 3S ↔ P 2S ↔ P 3S ↔ P 4S.

D: MF 3S ↔ AL′ 3S ↔ M 7S.

E: M 1S ↔ M 2S ↔ M 3S ↔ M 4S ↔ M 6S ↔ M 23S.

F: M 8S ↔ M 9S ↔ M 10S ↔ M 11S ↔ M 12S ↔ MR 14S(A)[1].

G: M 5S ↔ MR 14S(K).

In the following diagram when we write for example A → D we mean one of the equivalent propositions in A implies one of the equivalent propositions in D.

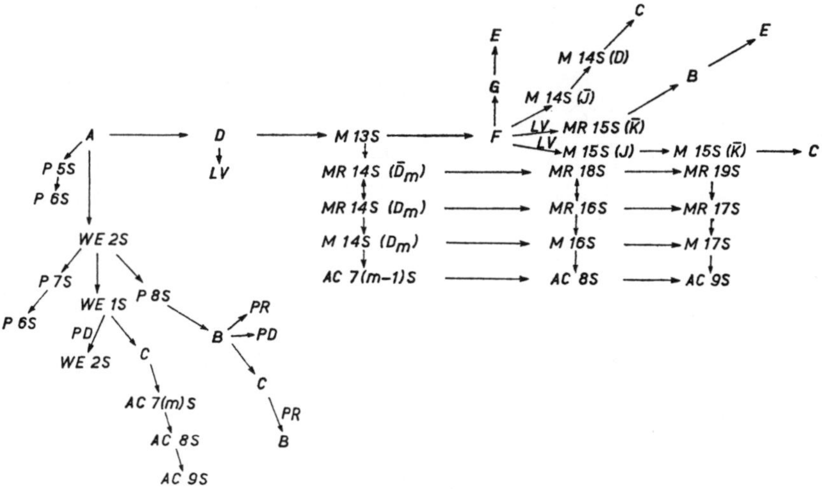

Using the axiom of regularity and that the class of individuals is a set it was shown that C → A, E → A, and AC 9S → A. Using the axiom of regularity and the existence of a relation which well-orders the class of individuals it was shown that P 6S → A. Hence, all the class forms are equivalent.

[1] Where $A = D, J, \bar{J}, U, \bar{U}, DK, DJ, DU, KJ, KU$, or JU.

LIST OF THE SET FORMS [1]

WE 1: Every set can be well-ordered. (1, 1)

WE 2: Every set is equivalent to an ordinal number. (1, 1)

WE 3: Every set is equivalent to a subset of an ordinal number. (1, 1)

Let m be a natural number, $m \geq 1$.

WE 4(m): For every set x there exists an ordinal number α and a function f defined on α such that $f(\beta) \preccurlyeq m$ for every $\beta < \alpha$ and $\bigcup_{\beta<\alpha} f(\beta) = x$. (Levy 1961; 1, 1)

WE 5: There exists a natural number $m \geq 1$ such that WE 4(m). (Levy 1961; 1, 1)

WE 6: For every set x there exists a natural number $m \geq 1$, an ordinal number α, and a function f defined on α such that $f(\beta) \preccurlyeq m$ for every $\beta < \alpha$ and $\bigcup_{\beta<\alpha} f(\beta) = x$. (Levy 1961, Rubin 1962; 1, 1)

AC 1: If s is a set of non-empty sets, there is a function f such that for every $x \in s$, $f(x) \in x$. (Zermelo 1904; 2, 5)

AC 2: If t is a disjoint set of non-empty sets there is a set c which consists of one and only one element from each set in t. (Russell 1906; 2, 5)

AC 3: For every function f there is a function g such that for

[1] In parentheses following each of the forms we shall give the name of the one who is given credit for the form (when applicable) and the date followed by the section and page number where it occurs in the text. Rubin 1963 refers to the present work.

every x, if $x \in \mathscr{D}(f)$ and, $f(x) \neq \Lambda$ then $g(x) \in f(x)$. (Zermelo 1904; 2, 5)

AC 4: For every relation r there is a function f such that $\mathscr{D}(f) = \mathscr{D}(r)$ and $f \subseteq r$. (Bernays 1941; 2, 5)

AC 5: For every function f there is a function g such that $\mathscr{D}(g) = \mathscr{R}(f)$ and for every $x \in \mathscr{D}(g)$, $f(g(x)) = x$. (Bernays 1941; 2, 5)

AC 6: The Cartesian product of a set of non-empty sets is non-empty. (Zermelo 1904; 2, 5)

Let m be a natural number, $m \geq 1$.

AC 7(m): If s is a set of non-empty sets, there is a function f such that for every $x \in s$, $f(x) \neq \Lambda$, $f(x) \subseteq x$, and $f(x) \preccurlyeq m$. (Levy 1961; 2, 5)

AC 8: There exists a natural number $m \geq 1$ such that AC 7(m). (Levy 1961; 2, 5)

AC 9: If s is a set of non-empty sets, then there is a natural number $m \geq 1$ and a function f such that for every $x \in s$, $f(x) \neq \Lambda$, $f(x) \subseteq x$, and $f(x) \preccurlyeq m$. (Levy 1961, Rubin 1963; 2, 5)

T: For all sets x and y either $x \preccurlyeq y$ or $y \preccurlyeq x$. (Hartogs 1915; 3, 9)

T′: For every two non-empty sets, there is mapping of one onto the other. (Lindenbaum 1926, Sierpinski 1948; 3, 10)

M 1: If R is a transitive relation on a non-empty set x and if every subset of x which is linearly ordered by R has an R-upper bound, then there is an R-maximal element in x. (Bourbaki 1939, Tukey 1940; 4, 12)

M 2: If R is a transitive relation on a non-empty set x and if every subset of x which is well-ordered by R has an R-upper bound, then there is an R-maximal element in x. (Szele 1950; 4, 12)

M 3: If every non-empty nest which is contained in a non-empty set x has its union in x then x has a maximal element. (Kuratowski 1922, Zorn 1935; 4, 12)

M 4: If every non-empty well-ordered nest contained in a non-empty set x has its union in x, then x has a maximal element. (Kuratowski 1922; 4, 12)

M 5: If R is a transitive relation on x, then there exists a maximal subset of x which is linearly ordered by R. (Hausdorff 1914; 4, 12)

M 6: For every set x there exists a maximal nest contained in x. (Hausdorff 1914; 4, 12)

M 7: For every set x and every property P of finite character there exists a maximal subset of x which has the property P. (Bourbaki 1939, Teichmüller 1939, Tukey 1940; 4, 13)

M′ 1: If R is a transitive relation on x and if every subset of x which is linearly ordered by R has an R-upper bound and if $y \in x$ then there is an R-maximal element $z \in x$ such that either $y R z$ or $y = z$. (4, 14)

M′ 2: If R is a transitive relation on x and if every subset of x which is well-ordered by R has an R-upper bound and if $y \in x$, then there is an R-maximal element $z \in x$ such that either $y R z$ or $y = z$. (4, 14)

M′ 3: If every non-empty nest which is contained in x has its union in x and if $y \in x$ then there is a maximal element $z \in x$ such that $y \subseteq z$. (4, 14)

M′ 4: If every non-empty well-ordered nest contained in x has its union in x and if $y \in x$, then there is a maximal element $z \in x$ such that $y \subseteq z$. (4, 14)

M′ 5: If R is a transitive relation on x and y is a subset of x which is linearly ordered by R then there exists a maximal subset z of x such that $y \subseteq z$. (4, 14)

M′ 6: If y is a nest contained in x then there exists a maximal nest z contained in x such that $y \subseteq z$. (4, 14)

M′ 7: For every set x and every property P of finite character and for every subset y of x such that $P[y]$ there is a maximal subset z of x such that $y \subseteq z$ and $P[z]$. (4, 15)

M 8: Let x be a set and R a relation then there is a maximal set $y \subseteq x$ such that $y \times y \subseteq R$. (Gottschalk 1952; 4, 15)

M 9: Let x be a set and R a relation then there is a maximal set $y \subseteq x$ such that $y \times y \subseteq R \cup R^{-1}$. (Wallace 1944; 4, 15)

M 10: Let x be a set and R a relation then there is a maximal set $y \subseteq x$ such that $y \times y \subseteq \bar{R} \cup \bar{R}^{-1}$. (Gottschalk 1952; 4, 15)

M 11: Let x be a set and R a relation, then there is a maximal set $y \subseteq x$ such that $y \times y \subseteq R \cup R^{-1} \cup I$. (Gottschalk 1952; 4, 15)

M 12: Let x be a set and R a relation then there is a maximal set $y \subseteq x$ such that $y \times y \subseteq \bar{R} \cup \bar{R}^{-1} \cup I$. (Gottschalk 1952, 4, 15)

M 13: Let x be a set and R an n-ary relation then there is a maximal set $y \subseteq x$ such that $y^n \subseteq R$. (Gottschalk 1952; 4, 15)

M 14(A): Every set has a maximal subset which has the property A, where $A = D$ (Vaught 1952; 4, 23), $A = \bar{D}, J, \bar{J}$ (Kurepa 1952; 4, 23) $A = D_m$ (Chang 1960, Levy 1961; 4, 25), $A = \bar{D}_m$, D or K or $\bar{J}$, D or $\bar{J}$ or K^*, $\bar{D}$ or J or $\bar{J}$ or K, $\bar{D}$ or J or K^*. (Rubin 1958 and 1963; 4, 26, 29–30)

M 15($\bar{K}$): Every set can be anti-symmetrically linearly ordered and every set has a maximal subset which has the property $\bar{K}$. (Kurepa 1952; 4, 24)

M 16: There exists a natural number $m \geq 2$ such that M 14(D_m). (Chang 1960, Levy 1961; 4, 26)

M 17: For every set x there exists a natural number $m \geq 2$ such that x has a maximal subset which has the property D_m. (Chang 1960, Levy 1961, Rubin 1963; 4, 26)

M 18: There exists a natural number $m \geq 2$ such that M 14($\bar{D}!_m$). (Rubin 1963; 4, 27)

M 19: For every set x, there exists a natural number $m \geq 2$ such that x has a maximal subset which has the property $\bar{D}!_m$. (Rubin 1963; 4, 27)

M′ 20: If x is an arbitrary set and $e \in x$ then there exists a maximal subset y of x such that $e \in y$ and y has the property D or $\bar{D}$ or J or $\bar{J}$ or K or K^*. (Rubin 1958; 4, 29)

M 21($A : B$): Any set which contains a maximal subset with the property A contains a maximal subset with the property B, where $A = D$ and $B = \bar{D}, J, \bar{J}$ or K; $A = \bar{D}$ and $B = D, \bar{J}$ or K; $A = J$ and $B = D, \bar{D}, \bar{J}$ or K; $A = \bar{J}$ and $B = \bar{D}, J$ or K;

$A = K$ and $B = D, \bar{D}, J$, or $\bar{J}$; $A = \bar{K}$ and $B = D, \bar{D}, J, \bar{J}$, or K. (Rubin 1960; 4, 30)

M 22($A:B$): Every set can be anti-symmetrically linearly ordered and any set which contains a maximal subset with the property A contains a maximal subset with the property B, where $A \neq B$ and $A = D, \bar{D}, J, \bar{J}, K$ or $\bar{K}$ and $B = D, \bar{D}, J, \bar{J}, K$, or $\bar{K}$. (Rubin 1960; 4, 31)

M 23: Every set x contains a maximal subset y such that for every $s, t \in y$, $s \neq t$, either $s \in t$ or $t \in s$. (Rubin 1963; 4, 36)

AL 1: Every lattice with a unit element and at least one other element contains a maximal proper ideal as a subset. (Scott 1953; 5, 39)

AL′ 1: Every proper ideal of a lattice with a unit element can be extended to a maximal proper ideal. (Scott 1953, 5, 39)

AL′ 2: If A is an algebraic system, B a subalgebra, and $a \in A$, but $a \notin B$ then there exists a maximal subalgebra which contains B as a subset, but does not contain a as an element. (Blair 1957; 5, 39)

AL′ 3: If x is an arbitrary set, C a finitary closure operator, P a property of finite character, and y a C-closed subset of x which has the property P, then there is a maximal C-closed subset of x which contains y as a subset and has the property P. (Rubin 1963; 5, 39)

AL′ 4: If x is an arbitrary set, $y \subseteq x$, and $R : y$, then there is a maximal subset z of x such that $y \subseteq z$ and $R : z$. (Beth 1953; 5, 41)

AL 5: If B is a Boolean algebra and $S \subseteq B$ such that $0 \notin S$, then there exists a maximal proper ideal disjoint from S. (Mrowka 1955; 5, 42)

AL 6: If $\mathscr{L} = \langle L, \{\vee, \wedge\}\rangle$ is a complete lattice, $a, b, c \in L$, $a \leq b$, c is compact, $c \leq b$ and $c \nleq a$, then there exists an element $m \in L$ such that $a \leq m \leq b$, $c \nleq m$ and if there is a $d \in L$ such that $m < d \leq b$ then $c \leq d$. (Diener 1956; 5, 42)

AL 7: Every subalgebra is the intersection of all meet-irreducible sub-algebras containing it as a subset. (McCoy 1938, Birkhoff and Frink 1948, Fuchs 1949, Schmidt 1953; 5, 43)

AL 8: Every proper subalgebra has a proper meet-irreducible subalgebra containing it as a subset. (Diener 1956; 5, 43)

AL 9: The set of absolutely dispensible elements is equal to the intersection of the set of all maximal proper subalgebras. (Frattini 1885, Neumann 1937, Schmidt 1953; 5, 43)

AL 10: For every element in a basis there exists a maximal subalgebra which does not contain it as an element. (Schmidt 1953; 5, 43)

AL 11: If A has a finite basis and $B \subset A$ is a subalgebra then there exists a maximal subalgebra which contains B as a subset. (Krull 1929, Schmidt 1953; 5, 43)

CN 1: $m \cdot n = m + n$. (Tarski 1924; 6, 52)

CN 2: There is a cardinal number n such that $m = n^2$. (Tarski and Rubin; 6, 52)

CN 3: $m = m^2$. (Tarski 1924; 6, 52)

CN 4: There is an ordinal number α such that for all transfinite cardinal numbers m there is no well-ordered by $<$ set M of cardinal numbers between m and m^2 such that $m \simeq \alpha$. (Tarski 1924, Rubin 1963; 6, 52)

CN 5: If $m^2 = n^2$ then $m = n$. (Tarski 1924; 6, 53)

CN 6: If $m < n$ and $p < q$ then $m + p < n + q$. (Tarski 1924; 6, 53)

CN 7: If $m < n$ and $p < q$ then $m \cdot p < n \cdot q$. (Tarski 1924; 6, 53)

CN 8: If $m + p < n + p$ then $m < n$. (Tarski 1924; 6, 53)

CN 9: If $m \cdot p < n \cdot p$ then $m < n$. (Tarski 1924; 6, 53)

CN 10: There is a cardinal number n such that $m < n$ and for every cardinal number p if $m < p$ then $n \leq p$. (Tarski 1954; 6, 53)

CN 11: If p covers n then either $m \cdot n = m \cdot p$ or $m \cdot p$ covers $m \cdot n$. (Sudan 1938; 6, 53)

CN 12: If $m < n$ then there is a p such that $n = m \cdot p$. (Tarski and Sierpinski; 6, 53)

CN 13: If $m < n$ then $n \div m$ exists. (Tarski and Sierpinski; 6, 53)

CN 14: If $m < n$ then $n \div m = n$. (Tarski and Sierpinski; 6, 53)

CN 15: If $m + p = m + q$ then either $p = q$ or $p \leq m$ and $q \leq m$. (Tarski 1926; 6, 53)

CN 16: If $m + m < m + n$ then $m < n$. (Tarski 1926; 6, 53)

CN 17: If $m < n$ then $n - m$ exists. (Tarski 1926, Sierpinski 1947; 6, 53)

CN 18: If $m < n$ then $n - m = n$. (Tarski 1926; 6, 53)

CN 19: If $p < n$ and $q < n$ then $p + q \neq n$. (Tarski 1926; 6, 53)

CN 20: If $p < n$ and $q < n$ then $p \cdot q \neq n$. (Sierpinski 1946; 6, 53)

CN 21: Either $m + n = m$ or $m + n = n$. (Lesniewski, Sierpinski 1947; 6, 53)

CN 22: Either $m \cdot n = m$ or $m \cdot n = n$. (Lesniewski, Sierpinski 1947; 6, 53)

Let m be a cardinal number, $m > 1$.

CN 23(m): If $p^m < q^m$ then $p < q$. (Tarski 1926; 6, 53)

CN 24: There is a cardinal number $n > 1$ such that for all cardinal numbers p and q there is a cardinal number m, $1 < m \leq n$ such that $p^m < q^m$ implies $p < q$. (Tarski 1926, Rubin 1963; 6, 54)

CN 25: If $m^p < m^q$ and $m \neq 0$ then $p < q$. (Tarski 1926; 6, 54)

CN 26: If the greatest lower bound $(m \frown n)$ and the least upper bound $(m \smile n)$ of m and n exist then $m \cdot n = (m \frown n) \cdot (m \smile n)$. (Rubin 1963; 6, 54)

P 1: For any non-empty set s there exists a set t such that for all x, $x \in t$ if and only if $x \subseteq t$ and $\bar{\bar{s}} \not\leq \bar{\bar{x}}$. (Tarski 1938, 7, 63)

P 2: For any non-empty set s there exists a set t such that $t \approx \{x \colon x \subseteq t \text{ and } \bar{\bar{s}} \not\leq \bar{\bar{x}}\}$. (Tarski 1938; 7, 63)

P 3: If t is an infinite set, m and n cardinal numbers such that $\bar{\bar{t}} = m$ and $n \leq m$, and $s = \{x \colon x \subseteq t \text{ and } n \not< \bar{\bar{x}}\}$ then $\bar{\bar{s}} = m^n$. (Tarski 1938; 7, 63)

P 4: If $x \prec \bigcup_{i=0}^{\infty} A_i$ then there exists a finite integer n such that $x \preccurlyeq \bigcup_{i=0}^{n} A_i$. (Tarski 1926 and 1948; 7, 67)

P 5: The product of compact spaces is compact in the product topology. (Tychonoff 1935, Kelley 1950; 7, 69)

P 6: A formula having a model in a set of cardinality n also has a model in a set of cardinality m if $\aleph_0 \leq m \leq n$. (Vaught 1956; 7, 69)

P 7: A formula having a model in a set of cardinality $\aleph_0$ also has a model in a set of any cardinality greater than $\aleph_0$. (Vaught 1956; 7, 69)

P 8: If Q is a set of formulas in which the set of individual constants has cardinality m and every finite subset of Q has a model, then Q has a model in a set whose cardinality is not greater than $m + \aleph_0$. (Vaught 1956; 7, 69)

P 9: If x is a non-empty set and A is a proper class then A can be mapped onto x. (von Neumann 1925; 7, 71)

P 10: If x is a non-empty set and A is a proper class then $x \prec A$. (von Neumann 1925; 7, 71)

P 11: Every class has a maximal subclass which has the property $\bar{D}$. (Kurepa 1952, Rubin 1963; 7, 72)

P 12: For every relation R whose domain is a set there is a function f such that $\mathscr{D}(f) = \mathscr{D}(R)$ and $f \subseteq R$. (Bernays 1941, Rubin 1963; 7, 73)

P 13: For every non-empty set A, for every function B whose domain is A, and for each $a \in A$, B_a is a set, and for every function X whose domain is $\dot{B} \circ \epsilon^{-1}$,

$$\bigcap_{a \in A} \bigcup_{b \in B_a} X_{a,b} = \bigcup_{f \in \times B_a} \bigcap_{a \in A} X_{a,f(a)}.$$

(Collins 1954, Rubin 1963; 7, 73)

P 13s: For every non-empty set A, for every class-valued function B, and for every function X whose domain is $B \circ \epsilon^{-1}$,

$$\bigcap_{a \in A} \bigcup_{b \in B[a]} X_{a,b} = \bigcup_{f \in \times B[a]} \bigcap_{a \in A} X_{a,f(a)}.$$

(Collins 1954, Rubin 1963; 7, 73)

P 14: For every non-empty set A, for every function B whose domain is A, and for each $a \in A$, B_a is a set,

$$\bigcap_{a \in A} \bigcup_{b \in B_a} b = \bigcup_{f \in \times B_a} \bigcap_{a \in A} f(a).$$

(Collins 1954, Rubin 1963; 7, 73)

P 14s: For every non-empty set A, for every class valued function B,

$$\bigcap_{a \in A} \bigcup_{b \in B[a]} b = \bigcup_{f \in \times B[a]} \bigcap_{a \in A} f(a).$$

(Collins 1954, Rubin 1963; 7, 73)

P 15: For every non-empty set A,

$$\bigcap_{a \in A} \bigcup_{b \in a} b = \bigcup_{f \in C(A)} \bigcap_{a \in A} f(a),$$

where $C(A)$ is the set of all choice functions on A. (Collins 1954; 7, 74)

P 16: Let K be an arbitrary set and let A and B be functions with domain K. If $A_k \prec B_k$ for all $k \in K$ then $\bigcup_{k \in K} A_k \prec \times_{k \in K} B_k$. (König 1905, Zermelo 1908; 7, 75)

P 17: The power set of a well-ordered set can be well-ordered. (Rubin 1960; 7, 77)

P 18: Every linearly ordered set can be well-ordered. (Rubin 1960; 7, 77)

P 19: Let R be a transitive, anti-symmetric relation on a non-empty set x. Consider the set $z_t = \{s: s \in x, t \in x, s \neq t \text{ and } t R s\}$. Suppose, for each $t \in x$, there is an R-minimal element $u \in z_t$ such that $u R s$ for each $s \in z_t$, $u \neq s$. Then there exists an R-linearly ordered subset of x with no R-upper bound. (Blair and Tomber 1960, Rubin 1960; 7, 78)

P 20: It is not the case that there exists a set x such that if $y = \mathscr{P}(x) \sim \{\Lambda\}$ and $F = \{f: f \text{ is a function mapping } y \text{ into } x\}$ then there exists a function g such that $\mathscr{D}(g) = F$, $\mathscr{R}(g) \subseteq y$, and $f(g(f)) \notin g(f)$ for all $f \in F$. (Tarski 1951; 7, 79)

LIST OF THE CLASS FORMS

WE 1S: There is a relation R such that every set is well-ordered by R. (1, 84)

WE 2S: There is a relation R such that every class is well-ordered by R. (1, 84)

WE 3S: There is a function F such that every set x is well-ordered by $F(x)$. (1, 84)

WE 4S: There is a relation R such that R well-orders V and every proper initial R-section of V is a set. (1, 84)

WE 5S: If X is a proper class then X is equivalent to On. (1, 84)
Let m be a natural number, $m \geq 1$.

WE 6(m)S: For every class X there exists a function F defined on On such that $F(\alpha) \leqslant m$ for every $\alpha \in On$ and $\bigcup_{\alpha \in On} F(\alpha) = X$. (1, 84)

WE 7S: There exists a natural number $m \geq 1$, such that WE 6(m)S. (1, 84)

WE 8S: For every class X there exists a natural number $m \geq 1$ and a function F defined on On such that $F(\alpha) \leqslant m$ for every $\alpha \in On$ and $\bigcup_{\alpha \in On} F(\alpha) = X$. (1, 84)

AC 1S: If S is a class of non-empty sets, there is a function F such that for each $x \in S$, $F(x) \in x$. (Gödel 1940; 2, 86)

AC 2S: If T is a disjoint class of non-empty sets, there is a class C which consists of one and only one element from each set in T. (2, 86)

AC 3S: For every function F there is a function G such that for every x if $x \in \mathscr{D}(f)$ and $F(x) \neq \Lambda$ then $G(x) \in F(x)$. (2, 86)

AC 4S: For every relation R there is a function F such that $\mathscr{D}(F) = \mathscr{D}(R)$ and $F \subseteq R$. (Bernays 1941; 2, 86)

AC 5S: For every function F there is a function G such that $\mathscr{D}(G) = \mathscr{R}(F)$ and for every $x \in \mathscr{D}(G)$, $F(G(x)) = x$. (Bernays 1941; 2, 86)

Let m be a natural number, $m \geq 1$.

AC 7(*m*)S: If S is a class of non-empty sets, there is a function F such that for every $x \in S$, $F(x) \neq \Lambda$, $F(x) \subseteq x$, and $F(x) \preccurlyeq m$. (2, 86)

AC 8S: There exists a natural number $m \geq 1$ such that AC 7(m)S. (2, 86)

AC 9S: Is S is a class of non-empty sets, then there is a natural number $m \geq 1$ and a function F such that for every $x \in S$, $F(x) \neq \Lambda$, $F(x) \subseteq x$, and $F(x) \preccurlyeq m$. (2, 86)

M 1S: If R is a transitive relation on a non-empty class X and if every subset of X which is linearly ordered by R has an R-upper bound then either there is an R-maximal element in X or X has a subclass which is a proper class and which is linearly ordered by R. (3, 89)

M 2S: If R is a transitive relation on a non-empty class X and if every subset of X which is well-ordered by R has an R-upper bound then either there is an R-maximal element in X or X has a subclass which is a proper class and which is well-ordered by R. (3, 89)

M 3S: If every non-empty nest which is a subset of a non-empty class X has its union in X, then either X has a maximal element or there is a subclass of X which is a nest and a proper class. (3, 89)

MF 3S: Let F be a class-valued function, Π a predicate such that for each class X

(i) $\Pi[X] \leftrightarrow (x)[x \subseteq X \rightarrow x \in \mathscr{D}(F)$ and $F[x] \subseteq X]$

(ii) $x \subseteq F[x]$ for all $x \in \mathscr{D}(F)$, and

(iii) if for every non-empty nest $Y \subseteq \mathscr{D}(F)$, $\Pi[\bigcup_{y \in Y} F[y]]$, then if there is a class which has the property Π, then there is a maximal class which has the property Π. (3, 91)

M4S: If every non-empty well-ordered nest which is a subset of a non-empty class X has its union in X, then either X has a maximal element or there is a subclass of X which is a well-ordered nest and a proper class. (3, 89)

M 5S: If R is a transitive relation on X, then there exists a maximal subclass of X which is linearly ordered by R. (3, 89)

M 6S: For every class X there exists a maximal nest contained in X. (3, 89)

M 7S: For every class X and every property P of finite character there exists a maximal subclass of X which has the property P. (3, 89)

M 8S: Let X be a class and R a relation; then there is a maximal class $Y \subseteq X$ such that $Y \times Y \subseteq R$. (3, 89)

M 9S: Let X be a class and R a relation; then there is a maximal class $Y \subseteq X$ such that $Y \times Y \subseteq R \cup R^{-1}$. (3, 89)

M 10S: Let X be a class and R a relation; then there is a maximal class $Y \subseteq X$ such that $Y \times Y \subseteq \bar{R} \cup \bar{R}^{-1}$. (3, 89)

M 11S: Let X be a class and R a relation; then there is a maximal class $Y \subseteq X$ such that $Y \times Y \subseteq R \cup R^{-1} \cup I$. (3, 89)

M 12S: Let X be a class and R a relation; then there is a maximal class $Y \subseteq X$ such that $Y \times Y \subseteq \bar{R} \cup \bar{R}^{-1} \cup I$. (3, 89)

M 13S: Let X be a class and R an n-ary relation; then there is a maximal class $Y \subseteq X$ such that $Y^n \subseteq R$. (3, 89)

M 14S(A): Every class contains a maximal subclass which has the property A, where $A = D$, $\bar{J}$, or D_m. (3, 89)

MR 14S(A): For every relation R and class X there exists a maximal subclass Y such that for all $x, y \in Y$, $x \neq y$, $R''\{x\} \, A \, R''\{y\}$, where $A = D, \bar{D}, J, \bar{J}, K, D_m, \bar{D}_m, U, \bar{U}, DK, DJ, DU, KJ, KU$, or JU. (3, 90)

M 15S(A): The universe can be anti-symmetrically linearly ordered and every class has a maximal subclass which has the property A, where $A = J$ or $\bar{K}$. (3, 90)

MR 15S($\bar{K}$): The universe can be anti-symmetrically linearly ordered and for every relation R and class X there exists a maximal subclass Y such that for all $x, y \in Y$, $x \neq y$, $R''\{x\} \bar{K} R''\{y\}$. (3, 91)

M 16S: There exists a natural number $m \geq 2$ such that M 14S(D_m). (3, 89)

MR 16S: There exists a natural number $m \geq 2$ such that MR 14S(D_m). (3, 91)

M 17S: For every class X there exists a natural number $m \geq 2$ such that X has a maximal subclass which has the property D_m. (3, 89)

MR 17S: For every relation R and class X there is a natural number $m \geq 2$ and a maximal subclass Y such that for all $y_1, y_2, \ldots, y_m \in Y$, y_i's distinct, $\bigcap_{i=1}^{m} R''\{y_i\} = \Lambda$. (3, 91)

MR 18S: There exists a natural number $m \geq 2$ such that MR 14($\bar{D}!_m$). (3, 91)

MR 19S: For every relation R and class X there exists a natural number $m \geq 2$ and a maximal subclass Y such that for all $y_1, y_2, \ldots, y_m \in Y$, $\bigcap_{i=1}^{m} R''\{y_i\} \neq \Lambda$. (3, 91)

M 23S: Every class X contains a maximal subclass Y such that for every $s, t \in Y$, $s \neq t$, either $s \in t$ or $t \in s$. (3, 89)

AL′ 3S: If X is an arbitrary class, C a finitary closure operator, P a property of finite character, and Y a C-closed subclass of X which satisfies P, then there is a maximal C-closed subclass of X which contains Y as a subclass and satisfies P. (3, 92)

P 1S: Every proper class can be mapped onto the universe. (von Neumann 1925; 4, 107)

P 2S: For every non-empty class A, for every function B whose domain is A, and for each $a \in A$, B_a is a set, and for each function X whose domain is $B \circ \epsilon^{-1}$,

$$x \in \bigcap_{a \in A} \bigcup_{b \in B_a} X_{a,b} \leftrightarrow (\exists F)[F \text{ is a function, } \mathscr{D}(F) = A,$$

$$(a)[a \in A \rightarrow F(a) \in B_a] \text{ and } (a)[a \in A \rightarrow x \in X_{a,F(a)}]] \ (4, 107)$$

P 2sS: For every non-empty class A, for every class-valued function B whose domain is A, and for each function X whose domain is $B \circ \epsilon^{-1}$,

$$x \in \bigcap_{a \in A} \bigcup_{b \in B[a]} X_{a,b} \leftrightarrow (\exists F)[F \text{ is a function, } \mathscr{D}(F) = A,$$
$$(a)[a \in A \rightarrow F(a) \in B[a]] \text{ and } (a)[a \in A \rightarrow x \in X_{a,F(a)}]]. \text{ (4, 107)}$$

P 3S: For every non-empty class A, for every function B whose domain is A, and for each $a \in A$, B_a is a set,

$$x \in \bigcap_{a \in A} \bigcup_{b \in B_a} b \leftrightarrow (\exists F)[F \text{ is a function, } \mathscr{D}(F) = A,$$
$$(a)[a \in A \rightarrow F(a) \in B_a], \text{ and } (a)[a \in A \rightarrow x \in F(a)]]. \text{ (4, 107)}$$

P 3sS: For every non-empty class A, for every class-valued function B whose domain is A,

$$x \in \bigcap_{a \in A} \bigcup_{b \in B[a]} b \leftrightarrow (\exists F)[F \text{ is a function, } \mathscr{D}(F) = A,$$
$$(a)[a \in A \rightarrow F(a) \in B[a]], \text{ and } (a)[a \in A \rightarrow x \in F(a)]]. \text{ (4, 107)}$$

P 4S: For every non-empty class A, $x \in \bigcap_{a \in A} \bigcup_{b \in a} b \leftrightarrow (\exists F)[F$ is a function, $\mathscr{D}(F) = A$, $(a)[a \in A \rightarrow F(a) \in a]$, and $(a)[a \in A \rightarrow x \in F(a)]$. (4, 107)

P 5S: For every proper class X, $\mathscr{P}(X) \leqslant X$. (4, 107)

P 6S: The power class of a well-ordered class can be well-ordered. (4, 107)

P 7S: Every linearly ordered class can be well-ordered. (4, 107)

P 8S: There is a propositional function φ such that

$$(\exists X)(Y)[Y \neq \Lambda \rightarrow (\iota Z)\varphi(Z, X, Y) \in Y].$$

(Hilbert 1923; 4, 108)

LIST OF FORMS WEAKER THAN THE AXIOM OF CHOICE

PR: For every relation R there exists a relation S such that $\mathscr{D}(R) = \mathscr{D}(S)$, $S \subseteq R$ and $S''\{x\}$ is a set. (2, 86)

PD: For every non-empty relation R, if $\mathscr{R}(R) \subseteq \mathscr{D}(R)$, then there is a function f, $\mathscr{D}(f) =$ set of natural numbers, such that for each natural number n, $f(n)\ R\ f(n+1)$. (Tarski 1948; 2, 88)

LV: V can be anti-symmetrically linearly ordered. (3, 101)

BIBLIOGRAPHY [1]

[1] BACHMANN, H. Transfinite Zahlen. (Ergebnisse der Math. und ihrer Grenzgebiete Heft 1) Springer-Verlag, Berlin, Göttingen and Heidelberg (1955) 204 pp.

[1] BERNAYS, P. A system of axiomatic set theory I. *J. of S. L.* **2** (1937) 65–77.

[2] II **6** (1941) 1–17.

[3] III **7** (1942) 65–89.

[4] IV **7** (1942) 133–145.

[5] V **8** (1943) 89–106.

[6] VI **13** (1948) 65–79.

[7] VII **19** (1954) 81–96.

[1] BETH, E. W. Some consequences of the theorem of Löwenheim-Skolem-Gödel-Malcev. Indag. Math. **15** (1953) 66–71.

[2] The foundations of mathematics. Studies in Logic, North-Holland Publishing Company, Amsterdam (1959) 741 pp.

[1] BIRKHOFF, G. Lattice theory. A. M. S. Colloq. Publ. 25, New York (1940) 155 pp.

[1] BIRKHOFF, G. and FRINK, O. Representations of lattices by sets. *Trans. A. M. S.* **64** (1948) 299–316.

[1] BLAIR, R. L. and TOMBER, M. L. The axiom of choice for finite sets. *Proc. A. M. S.* **11** (1960) 222–226.

[1] BOURBAKI, N. Théorie des ensembles (fascicule de résultats) Act. Sci. et Ind. 846, Hermann et Cie., Paris (1939). 2nd edition (1951) 50 pp.

[2] Sur le théorème de Zorn. *Archiv der Math.* **2** (1950) 434–437.

[1] BÜCHI, J. R. Representation of complete lattices by sets. *Portug. Math.* **11** (1952) 151–167.

[1] CHANG, C. C. Maximal *n*-disjointed sets and the axiom of choice. *Fund. Math.* **49** (1960) 11–14.

[1] COLLINS, G. E. Distributivity and an axiom of choice. *J. of S. L.* **19** (1954) 275–277.

[1] The abbreviations *J. of S. L.* for *Journal of Symbolic Logic* and *A. M. S.* for *American Mathematical Society* will be used throughout the bibliography.

[1] DIENER, K. H. Über zwei Birkhoff-Frinksche Struktursätze der Allgemeinen Algebra. *Archiv der Math.* **7** (1956) 339–345.

[1] FRAENKEL, A. Über den Begriff "definit" und die Unabhängigkeit des Auswahlaxioms. *Sitz. d. Preuss. Akad. d. Wiss.* (1922) 253–257.

[2] Über eine abgeschwächte Fassung des Auswahlaxioms. *J. of S. L.* **2** (1937) 1–25.

[1] FRAENKEL, A. and BAR-HILLEL, Y. Foundations of set theory. Studies in Logic, North-Holland Publishing Company, Amsterdam (1958) 415 pp.

[1] FRATTINI, G. Intorno alla generagione dei gruppi di operazioni. *Atti Acad. Lincei Rend.* **4** Ser. 1 (1885) 281–285.

[1] FRINK, O. A proof of the maximal chain theorem. *Amer. Jour. of Math.* **74** (1952) 676–678.

[1] FUCHS, L. Über die Ideale arithmetischen Ringe. *Comment. Math. Helvet.* **23** (1949) 334–341.

[1] GÖDEL, K. The consistency of the axiom of choice and of the generalized continuum-hypothesis with the axioms of set theory. Annals of Math. Studies, No. 3, Princeton (1940) 66 pp.

[1] GOTTSCHALK, W. H. The extremum law. *Proc. A. M. S.* **3** (1952) 631.

[1] HALPERN, J. D. The independence of the axiom of choice from the Boolean prime ideal theorem. *Notices A. M. S.* **8** (1961) 279–80 (Abstract).

[1] HARTOGS, F. Über das Problem der Wohlordnung. *Math. Annalen* **76** (1915) 436–443.

[1] HAUSDORFF, F. Grundzüge der Mengenlehre. W. de Gruyter & Co., Leipzig (1914) 476 pp.

[2] (2nd revised ed.) Berlin & Leipzig (1927) 285 pp.

[1] HENKEN, L. The completeness of the first order functional calculus. *J. of S. L.* **14** (1949) 159–166.

[1] HESSENBERG, M. Kettentheorie und Wohlordnung. *J. reine angew. Math.* **135** (1909) 81–133.

[1] HILBERT, D. Neubegründung der Mathematik. Erste Mitteilung. *Abh. Math. Sem. Univ. Hamburg* **1** (1922) 157–177.

[2] Die Logischen Grundlagen der Mathematik. *Math. Annalen* **88** (1923) 151–165.

[1] KELLEY, J. L. The Tychonoff product theorem implies the axiom of choice. *Fund. Math.* **37** (1950) 75–76.

[2] General Topology. D. von Nostrand, New York (1955) 298 pp.

[1] KNESER, H. Eine direkte Ableitung des Zornschen Lemmas aus dem Auswahlaxiom. *Math. Zeit.* **53** (1950) 110–115.

[1] KÖNIG, J. Zum Kontinuumproblem. *Math. Annalen* **60** (1905) 177–180.

[1] KRULL, W. Die Idealtheorie in Ringen ohne Endlichkeitsbedingungen. *Math. Annalen* **101** (1929) 729–744.

[1] KRUSE, A. H. Some developments in the theory of numerations. *Trans. A. M. S.* **97** (1960) 523–553.

[1] Kuratowski, C. Une méthode d'élimination des nombres transfinis des raisonnements mathématiques. *Fund. Math.* **3** (1922) 76–108.

[1] Kurepa, G. Sur la relation d'inclusion et l'axiome de choix de Zermelo. *Bull. Soc. Math. France* **80** (1952) 225–232.

[1] Levi, B. Intorno alla teoria degli aggregati. *Istituto Lombardo di Sc. e Lettere, Rendiconti* (2) **35** (1902) 863–868.

[1] Lindenbaum, A. and Mostowski, A. Über die Unabhängigkeit des Auswahlaxioms und einigen seiner Folgerungen. *C. R. Varsovie* **31** (1938) 27–32.

[1] Lindenbaum, A. and Tarski, A. Communication sur des recherches de la théorie des ensembles. *C. R. Varsovie* **19** (1926) 299–330.

[1] McCoy, N. H. Subrings of infinite direct sums. *Duke Math. J.* **4** (1938) 486–494.

[1] Malcev, A. Untersuchungen aus dem Gebiete der mathematischen Logik. *Mat. Sbornik*, N. S. **1** (1936) 323–336.

[1] Mendelson, E. Some proofs of independence in axiomatic set theory. *J. of S. L.* **21** (1956) 291–303.

[2] The independence of a weak axiom of choice. *J. of S. L.* **21** (1956) 350–366.

[1] Mirimanoff, D. Remarques sur la théorie des ensembles et les antinomies cantoriennes. *L'Ens. Math.* **19** (1917) 209–217.

[1] Moore, R. L. Foundations of point set theory. A. M. S. Colloq. Publ. **13**, New York (1932) 486 pp.

[1] Mostowski, A. Über die Unabhängigkeit des Wohlordnungssatzes vom Ordnungsprinzip. *Fund. Math.* **32** (1939) 201–252.

[2] Axiom of choice for finite sets. *Fund. Math.* **33** (1945) 137–168.

[3] On the principle of dependent choices. *Fund. Math.* **35** (1948) 127–130.

[1] Mrowka, S. On the ideals' extension theorem and its equivalence to the axiom of choice. *Fund. Math.* **43** (1955) 46–49.

[2] Two remarks on my paper "On the ideals' extension theorem and its equivalence to the axiom of choice". *Fund. Math.* **46** (1958) 165–166.

[1] Neumann, B. H. Some remarks on infinite groups. *J. London Math. Soc.* **12** (1937) 120–127.

[1] von Neumann, J. Eine Axiomatisierung der Mengenlehre. *J. reine angew. Math.* **154** (1925) 219–240.

[2] Die Axiomatisierung der Mengenlehre. *Math. Zeit.* **27** (1928) 669–752.

[3] Über eine Widerspruchsfreiheitsfrage in der axiomatischen Mengenlehre. *J. reine angew. Math.* **160** (1929) 227–241.

[1] Peano, G. Démonstration de l'intégrabilité des équations différentielles ordinaires. *Math. Annalen* **37** (1890) 182–229.

[1] Quine, W. v. O. Mathematical Logic. Revised edition. Harvard University Press, Cambridge, Mass. (1951) 346 pp.

[1] Robinson, R. M. The theory of classes. A modification of von Neumann's system. *J. of S. L.* **2** (1937) 29–36.

[1] Rubin, H. On a problem of Kurepa concerning the axiom of choice. *Notices A. M. S.* **5** (1958) 378 (Abstract).

[2] Two propositions equivalent to the axiom of choice only under both the axioms of extensionality and regularity. *Notices A. M. S.* **7** (1960) 381 (Abstract).

[1] Rubin, H. and Rubin, J. E. Some new forms of the axiom of choice. *Notices A. M. S.* **7** (1960) 380 (Abstract).

[1] Russell, B. On some difficulties in the theory of transfinite numbers and order types. *Proc. London Math. Soc.* (2) **4** (1906) 29–53.

[1] Schmidt, J. Einige grundlegende Begriffe und Sätze aus der Theorie der Hüllenoperatoren. Ber. Math. Tag., Berlin (1953) 21–48.

[1] Scott, D. The theorem on maximal ideals in lattices and the axiom of choice. *Bull. A. M. S.* **60** (1953) 83 (Abstract).

[2] Definitions by abstraction in axiomatic set theory. *Bull. A. M. S.* **61** (1955) 442 (Abstract).

[1] Schoenfield, J. R. The independence of the axiom of choice. *J. of S. L.* **20** (1955) 202 (Abstract).

[1] Sierpinski, W. Les exemples effectifs et l'axiome du choix. *Fund. Math.* **2** (1921) 112–118.

[2] Sur une proposition équivalente à l'axiome du choix. *Actas Acad. Ci. Lima* **11** (1946) 111–112.

[3] L'hypothèse généralisée du continu et l'axiome du choix. *Fund. Math.* **34** (1947) 1–5.

[4] Sur la différence de deux nombres cardinaux. *Fund. Math.* **34** (1947) 119–126.

[5] Sur une proposition de A. Lindenbaum équivalente à l'axiome du choix. *C. R. Varsovie* **40** (1948) 1–3.

[6] Leçons sur les nombres transfinis. Gauthier-Villars, Paris (1950) 240 pp.

[7] Cardinal and Ordinal numbers. Pań对stwome Wydawnictwo Naukowe, Warszawa (1958) 487 pp.

[1] Skolem, T. Über einige Grundlagenfragen der Mathematik. *Skrifter utgit av det Norske Vid.-Akad. i Oslo* **1** (1929) 1–49.

[1] Specker, E. The axiom of choice in Quine's "New foundations for mathematical logic". *Proc. Nat. Acad. Sci. U. S. A.* **39** (1953) 972–975.

[2] Zur Axiomatik der Mengenlehre (Fundierungs- und Auswahlaxiom). *Zeit. f. math. Logik u. Grundlagen d. Math.* **3** (1957) 173–210.

[1] Sudan, G. Sur une note de A. Tarski. *C. R. Rumania* **3** (1938) 7–8.

[1] Szele, T. On Zorn's lemma. *Public. Math. Debrecen* **1** (1950) 254–256.

[1] Szpilrajn-Marczewski, E. Sur deux propriétés des classes d'ensembles. *Fund. Math.* **33** (1945) 303–307.

[1] Tarski, A. Sur quelques théorèms qui équivalent à l'axiome du choix. *Fund. Math.* **5** (1924) 147–154.

[2] Fundamentale Begriffe der Methodologie der dedukitiven Wissenschaften. I. *Monatsh. Math. Phys.* **37** (1930) 360–404.

[3] Eine äquivalente Formulierung des Auswahlaxioms. *Fund. Math.* **30** (1938) 197–201.

[4] On well-ordered subsets of any set. *Fund. Math.* **32** (1939) 176–183.

[5] Axiomatic and algebraic aspects of two theorems on sums of cardinals. *Fund. Math.* **35** (1948) 79–104.

[6] Cancellation laws in the arithmetic of cardinals. *Fund. Math.* **36** (1949) 77–92.

[7] Cardinal algebras. Oxford University Press, New York (1949) 327 pp.

[8] On a statement related to the principle of choice. *Bull. A. M. S.* **57** (1951) 81 (Abstract).

[9] Theorems on the existence of successors of cardinals, and the axiom of choice. *Indag. Math.* **16** (1954) 26–32.

[10] The notion of rank in axiomatic set theory and some of its applications. *Bull. A. M. S.* **61** (1955) 443 (Abstract).

[1] TEICHMÜLLER, O. Braucht der Algebraiker das Auswahlaxiom? *Deutsche Math.* **4** (1939) 567–577.

[1] TUKEY, J. W. Convergence and uniformity in topology. Annals of Math. Studies, No. 2, Princeton (1940) 90 pp.

[1] TYCHONOFF, A. Über einen Functionenraüm. *Math. Annalen* **111** (1935) 762–766.

[1] VAUGHT, R. On the equivalence of the axiom of choice and a maximal principle. *Bull. A. M. S.* **58** (1952) 66 (Abstract).

[2] On the axiom of choice and some metamathematical theorems. *Bull. A. M. S.* **62** (1956) 262 (Abstract).

[1] WALLACE, A. D. A substitute for the axiom of choice. *Bull. A. M. S.* **50** (1944) 578.

[1] WESTON, J. D. A short proof of Zorn's lemma. *Archiv der Math.* **8** (1957) 279.

[1] WHITEHEAD, A. N. On cardinal numbers. *Amer. Journ. of Math.* **24** (1902) 367–394.

[1] ZERMELO, E. Beweis, das jede Menge wohlgeordnet werden kann. *Math. Annalen* **59** (1904) 514–516.

[2] Neuer Beweis für die Wohlordnung. *Math. Annalen* **65** (1908) 107–128.

[3] Untersuchungen über die Grundlagen der Mengenlehre. I. *Math. Annalen* **65** (1908) 261–281.

[1] ZORN, M. A remark on method in transfinite algebra. *Bull. A. M. S.* **41** (1935) 667–670.

INDEX

I. Special Letters and Symbols

II. Technical Terms